The Urban Book Series

The Urban Book Series is a resource for urban studies and geography research worldwide. It provides a unique and innovative resource for the latest developments in the field, nurturing a comprehensive and encompassing publication venue for urban studies, urban geography, planning and regional development.

The series publishes peer-reviewed volumes related to urbanization, sustainability, urban environments, sustainable urbanism, governance, globalization, urban and sustainable development, spatial and area studies, urban management, transport systems, urban infrastructure, urban dynamics, green cities and urban landscapes. It also invites research which documents urbanization processes and urban dynamics on a national, regional and local level, welcoming case studies, as well as comparative and applied research.

The series will appeal to urbanists, geographers, planners, engineers, architects, policy makers, and to all of those interested in a wide-ranging overview of contemporary urban studies and innovations in the field. It accepts monographs, edited volumes and textbooks.

Indexed by Scopus.

Nele Aernouts · Francesca Cognetti ·
Elena Maranghi
Editors

Urban Living Lab for Local Regeneration

Beyond Participation in Large-scale Social Housing Estates

Springer

Editors
Nele Aernouts
Department of Geography, Cosmopolis
Centre for Urban Research
Vrije Universiteit Brussel
Brussel, Belgium

Francesca Cognetti
Department of Architecture and Urban
Studies
Politecnico di Milano
Milano, Italy

Elena Maranghi
Milano, Italy

ISSN 2365-757X ISSN 2365-7588 (electronic)
The Urban Book Series
ISBN 978-3-031-19747-5 ISBN 978-3-031-19748-2 (eBook)
https://doi.org/10.1007/978-3-031-19748-2

This Springer imprint is published by the registered company Springer Nature Switzerland AG
The registered company address is: Gewerbestrasse 11, 6330 Cham, Switzerland

Funding Details

The research project that was at the base of this book—SoHoLab—was supported by ERA-NET COFUND Smart Urban Futures under Project No. 7102886.

A cofounding of LAB-U Laboratory of Urbanism, EPFL ENAC IA LAB-U and ERA-NET COFUND enabled to publish the book as Open Access.

The authors wish to thank the funding bodies for this support.

Contents

Abbreviations

ENoLL	European Network of Living Labs
ENSUF	ERA-NET Smart Urban Futures
ESF	European Social Fund
ULL	Urban Living Lab(s)
PRASS	Plan régional d'affectation du sol
SHP	Social housing provider

Chapter 1
Introduction: Framing Living Labs in Large-Scale Social Housing Estates in Europe

Nele Aernouts, Francesca Cognetti, and Elena Maranghi

Abstract Today, Living Labs are increasingly promoted as innovative tools to deal with urban regeneration in Europe. In this contribution, we look at their potential in the context of the regeneration of large-scale social housing estates. Starting from the results of the research project SoHoLab (2017–2020) and building on the contributions of this book, we identify Living Labs as practices that are at the margin of key regeneration processes and actors but that nonetheless play an important, enabling role in triggering a more broadly supported approach to regeneration. We use the metaphor of the 'interstice' to identify Living Labs' role of mediating across different social, institutional, disciplinary, departmental, and policy realms. Nevertheless, caution is warranted. Living Labs should not be considered *the* approach towards the urban regeneration of marginalized areas; their potential lies precisely in their hybrid and constantly transforming character. In order to steer regeneration practices and policies that are actually more inclusive, they should be accompanied by a critical and self-reflexive research attitude.

Keywords Living labs · Large-scale social housing estates · Interstice

1.1 Why Opt for a Living Lab Approach in Large-Scale Social Housing Estates?

In the often-cited *Dilemmas in a General Theory of Planning*, authors Rittel and Webber (1973) highlight the so-called wicked problems in an increasingly differentiated and pluralistic society, which cannot be tackled trough 'optimal solutions' by planners, not even through 'solutions'. According to them, 'social problems are never

N. Aernouts (✉) · E. Maranghi
Department of Geography, Cosmopolis Centre for Urban Research, Vrije Universiteit Brussel, Brussels, Belgium
e-mail: nele.aernouts@vub.be

F. Cognetti · E. Maranghi
Department of Architecture and Urban Studies DAStu, Politecnico di Milano, Milan, Italy
e-mail: francesca.cognetti@polimi.it

N. Aernouts et al. (eds.), *Urban Living Lab for Local Regeneration*,
The Urban Book Series, https://doi.org/10.1007/978-3-031-19748-2_1

solved. At best they are only re-solved—over and over again'. The most ambitious thing planners have left to do, they state, is 'to improve some characteristics of the world where people live'. These lines of thought immediately bring us to some challenges encountered in general theories of spatial planning regarding the regeneration of large-scale social housing estates.

The regeneration of large-scale social housing estates and the potential for mobilizing local knowledge and actors have received great attention in planning debates. In general, consensus exists on the need and importance of the involvement of inhabitants and other local stakeholders in planning processes. Nevertheless, doubts have been cast on the actual possibility of such involvement. France has probably the longest tradition of planning policies specifically focused on large-scale social housing estates. The *politique de la ville* has made the involvement of inhabitants one of its core principles for the rehabilitation of these neighbourhoods. According to studies on regeneration processes in this context (Donzelot 2013; Epstein 2014), one of the greatest obstacles to participation is the reluctance or inability of institutions and local politicians to share power with local lay people. Even if local people are well organized, they often remain 'on the margins of power' in participatory programmes (Carrel 2013; Van Beckhoven et al. 2005, p. 236). In that case, participation risks becoming a token, facilitating the approval of plans. Another frequently mentioned criticism is the difficulty of reaching a diversity of inhabitants in a de facto socially mixed population. It proves hard to include less-heard voices or to fully grasp points of view planners are not necessarily familiar with. This is related to the very nature of planning processes, which often fail to go beyond plans; quantitative data; and rational speech to include other types of expression, skills, and knowledge (Sandercock 2019). Finally, participation has been increasingly replaced by notions such as social activation and innovation, in which policymakers expect a more active and direct role of citizens in the transformation of their own environment (Cognetti, this volume). Such notions should be looked at with some suspicion in the case of large-scale social housing estates, where tenants often already deal with not feeling understood or heard, and where they often have been framed as passive recipients of urban welfare (Flint 2004). These different criticisms and sensibilities, which are in line with the very nature of politics and democracy, have made some authors speak of 'impossible participation' (Warin 1995).

Intertwined with these criticisms, theories of planning have been developed and promoted in relationship to the regeneration of large-scale social housing estates. For instance, Hall and Rowlands (2005) put forward the collaborative planning model by Healey (1997), who paid strong attention to the acquisition of various forms of knowledge through interdisciplinary research, to the relationship-building between inhabitants and other actors in the planning process, and to bridging policy domains. This model has been criticized for its focus on consensus and loose attention to power dimensions within planning processes. Arguably, this criticism might be more related to the fact that the complexity of the model wasn't fully captured. Nevertheless, over time new models have been created that instead rely on self-organized or antagonistic stances to planning (Aernouts et al. 2020b). Surely, the politicization of subjects through the development of counter-hegemonic narratives sounds more empowering than a focus on consensus-building. However, self-organization or political formation might not always be evident in places where people are often struggling to make ends meet or where they have little confidence that their neighbourhood will actually improve. More generally, these 'ideal types' of planning have not necessarily made it to the table of policymakers or to daily practices of regeneration.

Hence, rather than starting from universal models of planning, in this book, we engage with experiential and reflexive ways of learning and engaging with planning in such spaces. By taking a Living Lab approach to urban regeneration in high-rise estates, we value cautious and modest interventions that can positively inform planning processes in these areas and hopefully induce change with regard to which subjects are heard and seen.

1.2 The Context of This Book

This book and its attempt to identify the potential of Living Lab approaches in large-scale social housing estates arose from the research project 'The regeneration of large-scale social housing estates through Living Labs' (SoHoLab). In this project, research units from the cities of Brussels (Belgium), Milan (Italy), and Paris (France) each focused on a particular case study to examine how regeneration processes in large-scale social housing estates could be better attuned to the needs and concerns of local inhabitants and organizations. While 'better' in this sentence related to the actual belief that improvement is possible in highly institutionalized planning processes, the research aim was deliberately left open, as we were interested in various forms of involvement—from direct participation in planning processes to more indirect forms of knowledge acquisition—and in different methods of establishing these forms. In this process, we wanted to understand how we could more effectively open up the policy design and implementation phases, especially for local inhabitants of social housing estates. As such, we were close to 'action research' and 'engaged planning research', where planners are highly intentional about their interest and the kind of change they wish to promote. This implies being aware of the web of social relationships in the world we acted within, and thus of the challenges such change entails.

In order to comply with this objective, the research units first developed an analysis of the existing planning contexts (Aernouts et al. 2020a) and a review of the most relevant methodological tools related to the Urban Living Lab approach and participatory approaches that have been applied to large-scale social housing estates (Aernouts et al. 2020b; Lefrançois 2021). These explorations went hand in hand with a study and activities in a Living Lab in each case, where each case was in a different stage. The three cases were all large-scale social housing estates but were embedded in different national contexts in terms of welfare and planning systems. This offered a threefold demonstration of the possible employment of Living Labs.

The first Living Lab was built on the existing experience of the university lab Mapping San Siro. The San Siro neighbourhood is one of the largest public housing neighbourhoods of the city of Milan. It is not only characterized by a general lack of public intervention and investment but also by a wide spread of bottom-up initiatives promoted by local networks and inhabitants. Founded in 2013, Mapping San Siro was initiated with the aim of working within and together with the neighbourhood of San Siro in order to produce an experience based on knowledge-sharing between academia and civil society. This would be able to generate new representations of the neighbourhood and innovative ideas in order to promote change from within the neighbourhood. Since May 2014, the group has run a space in the neighbourhood thanks to an agreement with Aler (the regional agency for housing that owns and manages the neighbourhood) and the Lombardy Region in order to enhance processes of on-field coresearch, participatory planning, and networking between different local actors. The SoHoLab project helped foster and improve the existing university Lab by promoting pilot projects, stimulating dialogue and coresearch between the local and institutional levels, and stimulating and expanding self-reflection on the role of the Lab in and for the neighbourhood.

The aim of the second Lab, implemented in Greater Paris, was to nourish discussions in the two other contexts and better understand the effects of participation in the long run after the initial involvement. The French context is historically characterized by strong public interventions with regard to the renovation of large-scale social housing estates and the adoption of participatory tools. The research studied three rehabilitation projects in the Paris region, measuring the extent and forms of involvement of inhabitants after the design and achievement phase and reflecting on the device of participation. More specifically, it questioned whether their implication in sustainability issues changed the perception of inhabitants regarding their environment and their motivation towards achieving sustainable management. In addition to the retrospective study of former participatory approaches, a design studio was developed, reflecting on the tool of the Living Lab.

The third Living Lab experiment, in the Brussels Capital Region, consisted of a 3-year research engagement in the Peterbos neighbourhood, one of the largest social housing estates at the periphery of Brussels. The Brussels Capital Region is marked by relatively low levels of social housing (9%) but has developed several programmes to increase the offer and set high standards for renovation. The aim of the Living Lab was to bridge the gap between extra-local planning processes and social dynamics on the site in order to move towards an integrated approach to neighbourhood regeneration. The site under research was a high-rise social housing estate consisting of 1,400 apartments, managed by two social housing companies. The project aimed to gather in-depth knowledge on daily life in the neighbourhood, its management, and extra-local policies and measures in order to understand if and how these aspects can be better aligned. These ‘hidden’ layers of socio-spatial, urban, and institutional information were unravelled through an ethnographic research project on the site and on the spatial, institutional, and associative dynamics of the planning process and their respective impact on the site.

The discussions and exchanges between the different research units enabled the researchers to develop shared reflections regarding the promotion of Living Labs in the context of processes of regeneration in our cities. These outcomes were merged into a concluding report (Maranghi 2020) and into suggestions and recommendations for policymakers and practitioners involved in the regeneration of housing estates or other neighbourhoods (Ryckewaert 2020).

1.3 The Meaning and Potential of Living Labs in Social Housing Estates

Living Labs emerge in, among other things, EU programmes that aim for applied and policy-relevant research where an impact on-site or on the subject studied is expected during the course of the project. Hence, a direct (policy) impact is expected, as opposed to investments in 'fundamental' research, where separate trajectories are needed to implement solutions in wider society after the conclusion of the fundamental research stage. Such applied research usually enables the inclusion of a wide range of knowledge and action partners from private entities as well as civil society, next to academic research institutions.

While the inherent diversity and experimental nature of Living Labs makes it impossible to narrow them down to just one approach to urbanism, in this introduction, we would like to look at some dimensions Living Labs (could) address in the context of the regeneration of social housing estates. In this respect, there is one returning theme throughout the SoHoLab research and in the contributions in this book: the Living Lab as an 'interstice'. Of the various concepts used to grasp the Living Lab characteristics in our research and in this book—such as 'cross-boundary arena' (Concilio 2016, in Cognetti, this volume); enabling space (Cognetti, this volume); third place (Wachter, this volume); intermediary or third party (Boni, this volume); threshold space (Fava, this volume); interstitial or marginal space (Allemeersch, this volume); grey area, third place, or liminal space (Lefrançois, this volume); a 'layered professionality' (Grassi, this volume); and *mixité* (Vigano et al., this volume)—the 'interstice' most accurately grasps the inherent experimental and connective character of the Living Lab at different levels.

The notion of the 'interstice' is derived from anatomy, in which interstices are fluid-filled areas that surround cells or parts of an organ. They form a connective tissue that creates structural continuity between nervous systems and other tissues. Architects refer to them as leftover gaps between a building's walls, which are neither 'inside' nor 'outside' a building. In sociology, interstitial spaces are understood as spaces with a selective opening (Rémy 2015), where forms of 'micro-interaction' (Furnari 2014) or 'hybridization' happen that can help construct collective competences (Rémy 2015). The 'interstice' therefore seems an adequate metaphor for a practice that lies outside or at the margin of the direct workings of the system it interacts with but that nonetheless plays a substantial, enabling role in this system. Having discerned this common thread regarding the development of Living Labs in large-scale estates, the question remains as to which friction 'cells' and 'nervous systems' can be connected through the interstice. Based on the findings of this book and the SoHoLab project, we highlight three frictions where Living Labs can make a difference in the context of regeneration processes.

1.3.1 Inhabitants Versus Institutions: Creating Spaces for Micro-interaction

In places like large-scale social housing, the traditional relationships of political representation, the routine mechanisms of governance, and social cohesion are deeply compromised. Living Labs could work to 'mend' these relationships. They could function as metaphorical and physical platforms, where the interaction—even if conflictive—among different actors leads to the transformation of local practices of governance, spaces, relationships, etc. This is particularly relevant in contexts where institutions have progressively lost their connection with the areas in which they operate and hence the effectiveness of their activities. On the one hand, Living Labs could stimulate institutional learning (de Leonardis 2001; Donolo 1997) with regard to urban regeneration (Ostanel 2017). On the other hand, the learning process is wider and includes other actors that live and operate in the area, such as third-sector organizations, inhabitants, NGOs, etc. They can help foster interaction among institutions and local actors belonging to civil society, for instance, by bringing together different forms of knowledge, values, cultural belongings, and power positions.

In this sense, Living Labs could be, first, interpreted as relational fields, in which relationships of trust and cooperation can be strengthened and can enrich spatial capital. Second, they are fluid and incremental contexts within which actors can change their positions, roles, and points of view over time through different kinds of possible interactions (cooperative, conflictive, etc.). Through relationality, a 'mutual learning path [...] is generated through the recognition of a field where different voices and interests can be negotiated, where conflict can be seen not as a barrier' (Padovani 2016, p. 40) but as a first step towards communication. Third, they are contexts able to produce new knowledge as a result of the construction of bridges between social worlds. This aspect implies that all forms of knowledge are legitimate within the platform. Through the participation in these processes, actors are induced to negotiate their own values, roles, and understandings, and therefore they are enabled to build new ones, based on the relationships with other actors involved. What makes it possible to qualify Living Labs in this way is both the space given to mutual learning and the possibility to scale up and introduce these learnings in ordinary practices (both institutional and local ones).

As such, Living Labs can serve as local planning platforms in which it is not just possible to experiment with 'best practices' in the interaction between citizens and local organizations but also to effectively test new forms of governance and of collaboration between local institutions and the grassroots. Here, they can both contribute to a shifting governance culture at the institutional level (Mosseray et al.

2021) and enhance local consciousness at the neighbourhood level, essential for implementing more empowering regeneration processes that more accurately reflect local conditions and needs.

1.3.2 Social Versus Technical Disciplines and Departments: Bridging Approaches, Cultures, and Practices

The regeneration of contemporary cities cannot be limited to one discipline. In order to develop accurate understandings of ordinary life in our cities, and of large-scale social housing estates, in particular, disciplines such as architecture, planning, anthropology, and geography need to contaminate one another (Cellamare 2016; Cognetti and Fava 2017). This should not only be limited to triggering interdisciplinary approaches but also entail including knowledge of local inhabitants and workers in territorial analyses and planning approaches.

First, interaction and contamination among spatially oriented and socially oriented approaches is highly relevant, especially in sites where governance actions are designed mostly independently from one another. As Cognetti and Fava (2017) have pointed out, socially oriented and spatially oriented disciplines display different attitudes towards the city: the first are usually more effective when it comes to accessing a deeper understanding of certain phenomena, while the latter are more effective in producing activities and change. In the case of the SoHoLab project, the different research units combined these perspectives during the entire duration of the project. A long-lasting interdisciplinary collaboration allowed the units to reinforce the effectiveness of the research-action dimension. Second, Living Labs can be used to explore different languages and 'nonacademic' disciplines such as art, theatre, performance, video-making, narrative journalism, etc. These languages are especially important regarding the possibility of communicating more effectively with dwellers and with citizens more broadly, stimulating a deeper kind of participation. Such participation is not just related to the expression of an opinion but also to the emergence of personal stories, perceptions, feelings, etc., which are very important for understanding how space is lived (Rifaad and Aernouts 2022; Sandercock and Attili 2010). Third, Living Labs can combine 'expert' knowledge with local knowledge. This aspect is linked to the concept of 'situating' (Cognetti, this volume), which enables the researcher to engage in an intensive and long 'engaged ethnography' (Aernouts et al. 2020b) in the field. The researcher's contribution is crucial in different phases, from the understanding of the local context, to the implementation of activities, to the evaluation of research and activities promoted. In this respect, Lave and Wenger (1991) speak of 'situated learning': learning that takes place through relationships between people, which is connected to authentic, informal, and often unintended contextual learning (Maranghi, this volume).

The interaction among disciplines, nonacademic languages, and nonexpert knowledge not only enriches the action-research path but also helps enhance the understanding of the partiality of strictly disciplinary or sectorial points of view in respect of complex problems or situations (Maranghi, this volume). It opens up the planning field to uncertainty, doubt (Aernouts et al. 2020b), and self-reflection. It highlights the relevance of time associated with change and of ethical responsibility related to the presence—as researchers and practitioners—in the field. It is vital to maintain this kind of attitude, which associates a research-action framework (oriented to change and intervention) with deep self-reflection, which is especially and inherently part of the anthropological discipline.

1.3.3 Outside Versus Inside Views: Bringing Together Local Conceptions of and External Competency in Space

During the 1990s, many European large-scale social housing estates were subject to different programmes and policies aimed at their regeneration. Implemented urban policies included integrated physical and socio-economic interventions and acting in service of a wide and multiscalar change in the neighbourhood considered. After this period of integrated urban policies, resources currently allocated to the urban regeneration of large-scale social housing estates have become scarce in many areas. Often, different tools are applied independently from one another. Social housing estates are usually governed by multiple authorities who might each develop renovation and regeneration plans, social inclusion projects, and/or other actions and programmes in the neighbourhood. Typically, the renovation of homes is the responsibility of the local social housing landlord, while public space is managed and refurbished by the municipality. Separate renovation projects could target different buildings or even different aspects, such as the building envelope, the technical installations, or home interiors. Given the multiple actors, programmes, and tools, as well as the limited means to develop large-scale regeneration plans, coherence between these diverse types of intervention is often lacking.

These different interventions, each with their own temporality, create insecurity and uncertainty among inhabitants. For many local actors, it remains unclear which direction the transformation of the neighbourhood will take. Furthermore, this partial approach makes it more difficult to consider inhabitants' social practices, uses, and forms of appropriation that are often deemed inappropriate or illegal in a more positive light. With respect to this appropriation, the boundaries between public and private could be seen differently in light of the individual's desire to find a little privacy, both in- and outside the home. This implies that housing should be thought of beyond the physical boundaries of its walls, integrating 'secondary spaces' (Rémy 1999) in which individuals can give free rein to their need for intimacy and express their identity.

In that sense, a Living Lab can be the place where other ways of thinking about architecture and urban planning in relationship to the social housing question can be experimented with. In addition, a Living Lab can stimulate the development of a joint and overarching vision, or at least a framework that enables giving a perspective to inhabitants in order to situate different projects and identify priorities. Moreover, it is crucial to involve local actors and networks in the co-construction of this shared perspective concerning the most urgent field of intervention and the integration of different activities (social, economic, etc.). A shared vision furthermore allows to ensure connection and integration among different fields (spatial, social, economic, etc.), which is possible only when the Living Lab is able to involve different actors in the co-construction of a planning vision. Finally, in large-scale social housing estates that were designed and conceived under one master plan or a combination of complementary execution plans, it is crucial to safeguard the spatial and architectural coherence of the ensemble over time.

As was the case in the SoHoLab project, the frictions above also show the potential of the 'interstice' as a separate funding stream that is not financed by the partners directly involved in the regeneration process. The question remains then how such Living Lab practices can receive a more permanent anchoring or 'place', taking into account that their value lies in exactly their 'outside' or 'marginal' position.

1.4 The Book

This book tries to capture the nature of Living Labs in the regeneration of large-scale social housing estates as well as the practices in the field that can nourish and be nourished by new perspectives and tools developed by them. It is divided into three main chapters. It was developed on the basis of the SoHoLab project's reflection on Living Labs but also integrates perspectives and considerations produced by other researchers and professionals encountered by the SoHoLab team during the project or who were involved in the final cycle of online seminars organized in 2020 (SoHoLab 2020). This means that, in the spirit of defining the Living Lab as an open, interdisciplinary (or even transdisciplinary and not exclusively disciplinary), hybrid platform, not all contributions included refer explicitly to or have specifically used the Living Lab methodology. However, it is our opinion as editors that all contributions offer insights into and suggestions on how to adapt and shape the Living Lab, defined as in 'interstice' in the diverse meanings articulated above. Such a perspective helps one not to get 'trapped' in the tool but rather to employ it in relation to the characteristics of every single social housing estate.

The first chapter of this volume focuses on the Living Lab, exploring its potential for testing new forms of urban governance. The two other chapters open up different perspectives and methodologies that could contribute to and reinforce the adaptation of Living Labs in large-scale social housing estates. While the second chapter starts from a reflection on the research relationships with inhabitants, the last one moves to a reflection on the space of housing estates.

In her opening chapter, Francesca Cognetti argues that, under the right conditions, Living Labs can act as enabling and situated spaces counteracting the failure of certain participatory approaches to planning. She highlights that participation, and Living Labs in extension, has increasingly become a technical issue, often marked by pre-packaged formats and the support of technological tools. This entails the risk of downplaying its political significance. According to her, 'socially oriented' Living Labs can serve as platforms where processes of colearning take place and new forms of governance and interaction can be tested. As such, they can become an important field for defining and experimenting with new planning practices.

In his essay, Serge Wachter explores the darker side of Living Lab approaches. Based on a literature review of Urban Living Labs and EU guidelines regarding transition, he conceptualizes the Living Lab as a new model of governance used to experiment with sustainability solutions. As is the case with the concept of governance, this comes with several criticisms. In contrast to their high visibility and media coverage, Living Labs are often tools for tactical urbanism, developed at the margins of urban planning and only very moderately affecting the living conditions in our cities. In this sense, they do not address the radical urgency of climate change. When applied in so-called disadvantaged or sensitive areas, their impact might be stronger, yet can also be more disturbing. Indeed, according to Wachter, they are typical forms of remote governance, allocating duties and responsibilities to the private sector, local authorities, and/or underprivileged inhabitants. For the latter, Urban Living Labs can become expressions of governmentality, in which they are seemingly empowered to adopt goals of sustainability. As such, goals of social justice and sustainability are mixed with one another 'in a confusing fashion'. Should we even speak of 'bio-living labs', as Wachter boldly suggests? Be that as it may, the contribution reads as a warning for anyone involved in Living Labs to not uncritically adopt and impose the sustainability goals of the EU agenda.

In her contribution, Alice Boni offers some counterweight to the former contribution. She compares two experiences of 'local labs' in similar areas to discuss whether institutional leverage can help transfer Living Labs from a local to a regional or national level. According to the author, a policy transfer is possible, as long as there is a willingness on the part of the institutions to adopt a reflexive, open, and dialogical approach in the contexts in which they intervene.

Similarly, in the last contribution of the first part, Francesca Cognetti and Elena Maranghi discuss policy transfer in the context of Living Labs, but to the local level instead. The contribution investigates how a Living Lab can promote a more effective inclusion of inhabitants and local groups in urban policies. The authors underline the role of codesign as an interesting tool to not only promote urban change but also stimulate the empowerment of individuals and local communities.

The second part of the book is dedicated to reflecting on the social relations that are researched, built, and dissolved as part of research and action in large-scale social housing estates. Here, it is discussed what Living Labs can and do entail, starting from a perspective developed during a long-term stay in the neighbourhood. Researchers, artists, architects, and urbanists engage in this context and try to understand the environment of others. Although, as shown, ethnography and anthropology might be disciplines that come close to adopting an 'internal perspective' (Grassi, this volume), it remains impossible to assign it to individuals or to a single method. It is rather socially constructed, apprehended, invented, and carried out collaboratively (Fava, this volume). Therefore, researchers are bound to both remain with and release their disciplinary background and to attempt to inhabit the research relationships and spaces they aim to understand (Fava, this volume).

In the introduction of this second part, Ferdinando Fava discusses how the anthropological tradition can help us understand the relevance of interactions within Living Labs in large-scale social housing estates from an epistemological point of view. According to Fava, 'social bonds' should be the primary characteristic of a Living Lab. The potential space of the anthropologist, and the Living Lab in extension, is the threshold between in- and outsider that makes understanding through dialogical construction possible by virtue of an 'in-betweenness'. Inhabiting a place, through anthropology or by occupying a physical space in a social housing estate (as in the case of the SoHoLab) then becomes synonymous with inhabiting a research relationship and with trying to establish social bonds with concrete people that make up the place.

In this respect, Paolo Grassi, in the second chapter of this part, speaks about the frustrations that accompany such research relationships. According to Grassi, loving, hating, and failing are inextricably intertwined with fieldwork. Here, he draws a parallel between ethnographic fieldwork and the experimental character of Living Labs in planning. He denounces the rhetoric of success and empowerment within planning literature and practice. According to him, similar to Living Lab experiments applied in the physical sciences, remaining in place and accepting failures and frustrations are key. Indeed, they force planners to come up with newer and better explanations for the same phenomenon over and over again.

A third author who attempts to capture his own role in the context of the transformation of a social estate is Simon Allemeersch. He describes his experience in an atelier developed in an empty apartment of a high-rise social housing estate subject to demolition in Ghent (Belgium). According to him, the estate could be considered a 'broken institution', where the necessary formal order left the building. In this context, his atelier functioned as an interstitial space, not only welcoming inhabitants but also all kinds of actors, students, journalists, and social workers interested in the space. Here, the formal and informal, in- and outside came together and communicated with one another, helping him to decipher both.

In her contribution, Elena Maranghi puts social professionals and activists working in the territory of San Siro centre stage. She explains how the Mapping San Siro project, and later on the SoHoLab project in the Milan neighbourhood, stimulated the formation of a network among these actors. Over time, the lab helped them overcome conflicts and collectively build up knowledge. It pushed them to become a 'community of planning', strengthening their voice and agency with regard to local policies.

While the second part focuses on the relationships between 'people' developed as part of a Living Lab, the third part focuses on the characteristics of 'space' in large-scale social housing estates. It proposes possible forms of analysis and interventions that aspire to be capable of integrating 'the spatial' and 'the social'. Developed within the rationalist planning paradigm of modernist and functionalist urban design, following the principles of the Athens Charter (Mumford 2000), large-scale social housing estates have specific spatial forms that often makes them easily distinguishable from other neighbourhoods, especially medieval cities or city extensions dominated by nineteenth- and early twentieth-century development paradigms. This spatial form has been subject to lengthy debates and interpretations in architectural and planning theory. Yet, in the meanwhile, they have been inhabited and have played an important role in guaranteeing the right to housing for a very long time. Highbrow discussions regarding the adequateness of the spatial form have therefore become partially irrelevant. In this part, we want to highlight that Living Labs can play a role, capturing or informing various interpretations and understandings of the use of space.

In the first contribution to this third part, Lucia Capanema revisits the notion of the 'ghetto' in relationship to mobility, understood as a combination of motility, accessibility, and porosity. In the ghetto, immigrants and marginal populations are said to be stuck, as opposed to elites, who have the right to privileged centres. While a quantitative survey in the neighbourhood of San Siro, Milan, shows that inhabitants of these estates feel reasonably good in their neighbourhood and feel relatively well served by public transportation, a qualitative study highlights that mobility problems can be deeper and are not necessarily visible. As such, the author doubts whether inhabitants of social housing estates can enjoy destinations 'in their fullness'. An analysis of mobility, access, and porosity can inform new initiatives and interpretations of the meaning and value of space in these estates.

In their presentation of their analysis and master plan of the high-rise estate of Peterbos, Brussels (Belgium), Paola Vigano, Betrand Plewinski, Guillaume Vanneste, and Nicolas Willemet also do not shy away from interpreting large-scale social housing estates as 'enclaves for the poorest' or as spaces where 'the misery in the world' is concentrated. The bold expressions used for capturing large-scale social estates are somewhat in contrast to their sensitive reading of the hilly landscape, mundane infrastructures, and housing patrimony in the area. They do highlight the necessity for accompanying physical interventions by other forms of reform, such as social and economic interventions. They put forward four themes that connect both: a reinterpretation of modern heritage; the 'project of ground', activating ground floor spaces; a societal and ecological transition and public space as a space of emancipation and diversity. Building on these themes through 'the urban project', the authors offer new imaginaries to the area.

Contrary to these conceptual and political readings of the areas, the last two contributions start from an ethnographic analysis of the use of public spaces and shops in high-rise estates. In the third contribution, Jeanne Mosseray and Nele Aernouts highlight that in contemporary urbanism, and in European city renewal programmes such as the Neighbourhood Contracts more specifically, spatial interventions are often justified through social intentions, relying on older conceptions of community and social cohesion. The authors deconstruct such goals in the redesign of the ground floors in the same high-rise estate of Peterbos. Based on an ethnographic analysis of the actual use, occupation, and appropriation of the existing shops, they question the aim of architects, urbanists, social workers, and administrators to activate the ground floors in order to make the neighbourhood more vibrant. According to them, an ethnographic reading could help counter notions such as 'activation' and 'appropriation' and contribute to make more informed decisions on spaces.

By studying the use of public spaces in high-rise estates in the city of Fresnes, close to Paris, in the fourth contribution of this last part, Dominique Lefrançois criticizes participatory planning processes for turning a 'blind eye' to actually existing social practices that somehow already 'shape' large-scale social housing estates. She argues that these do not succeed in grasping other attempts of inhabitants to express opinions and viewpoints, for instance, by sending letters, signing petitions, and writing to the press. She questions whether, rather than focusing on inhabitants, Living Labs should instead focus on those actors conceiving and managing these spaces in order to improve the quality of their service. She furthermore criticizes the imposition of ecological imperatives on the areas examined, which are not in line with actual sustainable practices in the area and do not make an effort to start from actual practices in the area. She argues that in order to enable Living Labs to put themselves forward as spaces for ecological experimentation, challenging accepted ideas and offering visions going beyond the city/nature opposition, basic principles of ethnology should be taught in architecture schools.

1.5 Conclusion

The chapters expose strengths and weaknesses of the notion of the Living Lab and its implications when upscaling. Underlying scepticism about the notion is apparent, especially when seen in the light of European policies, subsidies, and funding programmes calling for the mobilization of citizens and local authorities in relationship to the adaptation of cities and regions to climate change. Here, the Living Lab becomes synonymous with the EU's entrepreneurial approach to the urgency, where local authorities and inhabitants are tempted to adopt 'innovative' strategies and answers to problems in order to increase their competitiveness and international appeal.

If anything, the added value of the Living Lab approach in planning processes seems more apparent when it is critically adopted by interdisciplinary research teams. Here, Living Labs can enable the bridging of knowledge and different social and professional worlds. They can help call on planning institutions to invest in large-scale social housing estates or encourage local voices to emerge and to be heard. A long-term presence and an embeddedness in the existing network of planning and local actors is key to their success. This shows that EU policies also have the potential to be 'appropriated' to locally steer more inclusive planning processes, especially in the highly institutionalized contexts of social housing estates. However, in order for a more permanent anchoring or 'place-taking' to happen, noninstrumental and open research trajectories in urban regeneration projects should be supported locally, beyond and in addition to the contracting out of master planning and participation trajectories to professionals.

In this introduction, we have argued that the innovative and experimental approach of Living Labs might help develop more fine-tuned answers to 'wicked' problems in urban planning. Building on the contributions in this volume, we suggest that 'innovation' and 'experimentation' should go along with a careful reading of large-scale social housing estates and with strengthening the accountability of those responsible for their management and renovation.

References

Aernouts N, Maranghi E, Ryckewaert M (eds) (2020a) The regeneration of large-scale Social Housing estates. Spatial, territorial, institutional and planning dimensions. Soholab, Brussels

Aernouts N, Maranghi E, Ryckewaert M (Eds) (2020b) Towards a definition of socially oriented Living Labs. Bussels: SoHoLab

Carrel M (2013) Faire participer les habitants? Citoyenneté et pouvoir d'agir dans les quartiers populaires. ENS Editions

Cellamare C (2016) Leggere l'abitare attraverso l'interdisciplinarietà e la ricerca-azione. Territorio 78:28–39

Cognetti F, Fava F (2017) La città indisciplinata. Note per un'agenda di ricerca. Tracce Urbane 1:126–136

De Leonardis O (2001) Fighting social exclusion: Between economic development and social wellbeing. Università degli Studi di Milano-Bicocca

Donolo C (1997) L'intelligenza delle istituzioni. Feltrinelli

Donzelot J (2013) La France des cités. Le chantier de la citoyenneté urbaine. Editions Fayard

Epstein R (2014) La nouvelle politique de la ville au prisme des évaluations du passé. Halsh-01023283

Flint J (2004) Reconfiguring agency and responsibility in the governance of social housing in Scotland. Urban Studies 41(1):151–172

Furnari S (2014) Interstitial spaces: Microinteraction settings and the genesis of new practices between institutional fields. Acad Manag Rev 39:439–462

Hall S, Rowlands R (2005) Place-making and large estates: theory and practice. In: van Kempen R, Dekker K, Hall S, Tosics I (eds) Restructuring large housing estates in Europe. Policy Press, University of Bristol, pp 47–62

Healey P (1997) Collaborative planning: shaping places in fragmented societies. Macmillan Press, London, England

Lave J, Wenger E (1991) Situated learning. Cambridge University Press, Legitimate peripheral participation

Lefrançois D (2021) From participation to Living Lab. A look back at experimental approaches to refurbishment, multi-partner projects and participatory initiatives. SoHoLab

Maranghi E, Cognetti F, Ryckewaert M, Aernouts N, Mosseray J, Lefrançois D, Wachter S, Labied N (2020) 5 themes from urban living lab experiences in large-scale social housing estates. Brussels: Soholab. http://www.soholab.org/publications/themes-from-urban-living-lab-experiences-in-large-scale-social-housing-estates

Mosseray J, Aernouts N, Ryckewaert M (2021) Institutional ethnography: a transformative mode of inquiry in the renovation of a Brussels high-rise housing estate. Europ Plann Stud. Advance online publication. https://doi.org/10.1080/09654313.2022.2057186

Mumford E (2000) The CIAM discourse on urbanism, 1928–1960. MIT Press

Ostanel E (2017) Urban regeneration and social innovation: The role of community-based organisations in the railway station area in Padua, Italy. J Urban Regeneration Renew 11(1):79–91
Padovani L (2016) Partecipazione come forma di apprendimento. Sentieri Urbani 7(21):39–43
Rémy J (1999) Le statut de l'espace dans les sciences sociales. Espaces Et Sociétés 96:165–220
Rémy J (2015) L'espace, un objet central de la sociologie. Eres
Rifaad Y, Aernouts N (2022) Storytelling as counternarrative: evidence from a Brussels social high-rise estate. Manuscript submitted for publication
Rittel HWJ, Webber MM (1973) Dilemmas in a general theory of planning. Policy Science 4:155–169
Ryckewaert M, Aernouts N, Cognetti F, Lefrançois D, Maranghi E, Mosseray J (2020) SoHoLab guidelines for the regeneration of large-scale social estates. Brussels: Soholab. http://www.soholab.org/publications/soholab-guidelines-for-the-regeneration-of-large-scale-social-estates
Sandercock L (2019) Making the invisible visible: a multicultural planning history. University of California Press. (Original work published 1998)
Sandercock L, Attili G (2010) Digital ethnography as planning praxis: an experiment with film as social research, community engagement and policy dialogue. Plan Theory Pract 11:23–45
SoHoLab (2020) SoHoLab webinars. http://www.soholab.org/news/soholab-webinars
Van Beckhoven E, van Boxmeer B, Garcia Ferrando L (2005) Local participation in Spain and the Netherlands. In: van Kempen R, Dekker K, Hall S, Tosics I (eds) Restructuring large housing estates in Europe. Policy Press, University of Bristol, pp 231–256
Warin P (1995) Les HLM: Impossible participation des habitants. In: Sociologie du travail, 37, 2, 151–176

Part I
A Critical Overview on Urban Living Labs in Large-Scale Social Housing Estates

Chapter 2
Beyond a Buzzword: Situated Participation Through Socially Oriented Urban Living Labs

Francesca Cognetti

Abstract In the broader framework of Living Labs and participatory planning, the essay proposes socially-oriented Urban Living Labs (ULLs) as a possible way of understanding and experimenting with participation in marginalized contexts. It does so by applying a focus on individual/collective capacities and enabling processes to support them. Drawing on the literature and the observation and implementation of concrete cases, the essay proposes a reflection on ULLs as situated environments in which "everyone's" capacities are formed and tested, thus challenging the functioning of local democracy. This implies a focus as much on residents and local agents as on institutions. The essay proposes a shift from Living Labs to socially oriented Urban Living Labs, in order to foster the social dimension of planning, questioning the mechanisms of involvement and support of the most fragile profiles, often excluded from the political process. The perspective is the implementation of an enabling and mutual learning process through devices to reinforce organizations and people's ability to reflect on, aspire to, and take action for the transformation of their life context, becoming real agents of change.

Keywords Living Labs · Participation · Situated approach · Marginalized contexts

2.1 Introduction

This essay proposes to analyze Urban Living Labs (ULLs) from the perspective of participatory planning. In the last decade, the need to adopt new forms of city governance seems increasingly evident, including various ways of interacting with a plurality of actors. Terms such as *inclusion*, *partnership*, *collaboration*, and *participation* characterize the rhetoric of different projects and policies, not only in the field of urban transformation but also in the fields of social cohesion and in cultural and integration projects. With respect to this important change, which marks an openness to these issues but also the risk of their uncritical and undifferentiated use (Brownill

F. Cognetti (✉)
Department of Architecture and Urban Studies DAStu, Politecnico di Milano, Milan, Italy
e-mail: francesca.cognetti@polimi.it

N. Aernouts et al. (eds.), *Urban Living Lab for Local Regeneration*,
The Urban Book Series, https://doi.org/10.1007/978-3-031-19748-2_2

and Parker, Plan Pract Res 25:275–282, 2010; Legacy, Plan theory 16:425–442, 2017), this essay reflects on the form and quality of the participation, of which the tools adopted through the ULLs are bearers, particularly in marginalized fields. In fact, introducing positive participatory processes into these contexts means centering concrete tools to open up the decision-making arena and also centering empowerment mechanisms and inclusive tools of co-learning and co-design. ULLs, under certain conditions, can be a device to strengthen this perspective, marking some innovations with respect to the now consolidated field of participation as opportunities 'in and for planning' (Concilio and Rizzo (eds) Rethinking the interplay between design and planning. Springer, 2016).

2.2 Which Participation and Why?

The last 20 years have been marked by experimentation with, dissemination of, and subsequent consolidation of practices of a participatory nature, even within the tools of territorial government at different scales, from urban regeneration to strategic planning. This is acknowledged by some studies in the field of urban planning in Europe as well.[1] In its gradual consolidation, the field of participation is in danger of losing some of its initial charge, which was aimed at the possibility of radically generating more open and inclusive projects. On the one hand, the participatory approach has represented a discontinuity with respect to the more traditional orientations of urban planning, marking the possibility of opening up the context—considered for a long time to be exclusively technical and presided over by experts—to broader political practices and citizenship. On the other hand, a certain pervasiveness of participatory tools and methods, in the face of not being as valid in terms of shared processes, has begun to raise doubts about the effectiveness of the approach and the possibility of its widespread use. Arguably, a participatory approach oriented toward urban transformation can be summarized as follows: participation is a process promoted by public actors to stimulate individual citizens and organizations to 'participate' in the material and immaterial transformations of the city, using techniques of shared decision-making. This statement contains the contours and the premises of an idea of participation, on which it is useful to reflect.

The assumptions seem to be the following:

- With respect to the participation process, the public entity is the promoter and has an active role: it establishes the framework for and the rules of involvement; it designs the political and project arena; it 'educates' the stakeholders, and it convenes the citizens;

[1] For an overview, see Nadin et al. (2021). The article explores the evolution of European spatial planning systems in their capacity to promote integration between policy sectors, to respond adaptively to changing societal and political conditions, and to involve and engage citizens in decision-making processes. Indeed, the reach and impact of these changes is still an open question.

- The outcome of participation consists of choices and directions that can guide, in a more or less binding way, transformations or lines of conduct in the future;
- Those who participate have the possibility to actually do so in terms of time and ability, making available their ideas and their skills to the collective process free of charge and freely;
- Third figures or facilitators of participation have a neutral role and simplify the exchange through the use of tools and techniques.

A wide range of research and practices look at participation, and many have been critical of the benefits of participation in planning and more broadly in local governance, underlining the 'crisis of participatory planning'.

Many authors have questioned whether these processes truly generate more open and inclusive policies, projects, and contexts. There are many perspectives critical of participation: some call it a 'nightmare' (Miessen 2010), others a 'tyranny' (Cooke and Kothari 2001), or something to be 'looked upon with suspicion' (Bobbio 2016) that is weakly oriented toward 'increas[ing] social justice' (Fung 2017). Participatory planning is in danger of becoming a pacified process, and the interaction between participants is more and more a functional moment to make decisions 'a little more shared' and to build consensus (Crosta 2003). This is even more plausible in marginal contexts, in which there are few experiments underway and in which the role of residents and local representatives is weak, while the presence of institutions is ambiguous and controversial. Besides, participation in the renovation of large-scale social-housing estates has been rather ineffective in the previous decades, primarily involving segments of the population, namely those who are best capable and most comfortable expressing themselves (Maranghi et al. 2020).[2] And in many cases, encouraging greater participation by marginalized groups does not safeguard values of equality, nor should it be presumed to do so (Beebeejaun 2006). For these reasons, participation could be considered an alternative among many possible policy tools, to be adopted only after a careful evaluation of its costs and benefits (Mela 2016). Moreover, participation can be articulated in different ways, depending on the territorial context, the political environment, and available instruments.

The aim of this essay is to investigate whether and under which conditions Living Labs can contribute to redefining and experimenting with participation in planning, especially in marginalized contexts. The objectives are therefore several:

- to insert the Living Lab tool into a broader framework of perspectives on participation, which problematize the relationship between technique, politics, and policies, providing a different understanding of the role of actors and the role of spaces in the interaction process;
- to reflect on the specificities of social ULLs as a possible way of understanding participation in marginal contexts from the perspective of capacities and skills,

[2] Mainly the voices of elderly retirees and what can be called the middle-class segment of the population are heard, at the expense of the voices of others such as young people, immigrant populations, etc. The risk is that this 'selective' participation and the dominant voice of some groups contributes to and increases, rather than regulates conflicts between generations, between long-term residents in the neighborhoods and the newly arrived (Lefrançois 2021).

and as a specific variation within a broader framework of experiments on Living Labs;
- to investigate the characteristics of social ULLs starting from an interdisciplinary and inter-scalar experiment called the SoHoLab project, where we had the opportunity to field-test the implementation of three Living Labs in different marginalized social-housing neighborhoods.[3] Through this empirical study, reflections emerge on how to understand participation in this type of context.

2.3 Three Key Perspectives on Democracy in Participatory Planning

When discussing participation, there are many aspects to pay attention to, carefully evaluating the costs and benefits of this way of enlarging decision-making processes and transformation of the city. The issue we are reflecting on concerns on the one hand the relationship between technical and political dimensions, and on the other the relationship between institutions and citizens (both understood as collaborative or autonomous promoters of participation). As we will see later, this aspect will also concern Living Labs, which, as tools for initiating and implementing inclusive processes, can have different characteristics, depending on how participation is understood.

In fact, participatory processes are often managed with sophisticated techniques of communication and dissemination, the collection of opinions, and the construction of forums and discussions, most often using pre-packaged formats with the support of technological tools. Living Labs have increasingly become platforms to support this approach.

However, the more participation is understood as a policy instrument that is technical in nature and therefore, to some extent, certain and adjustable, the more there is a risk of downplaying its political significance. This risk relates to power relations and their effects on people and places; the explicit confrontation between different parties and opinions, which can also result in conflict; the interdependencies between local conditions and urban or supra-local balances; and the structural factors that run through groups and society. Ultimately, the emphasis on the political disposition of participation refers to forms of democracy and the possibility of constructing frames of activation and confrontation in which an unpredictable and unexpected creative potential can be released, building new conditions for institutional and social action.

Contemporary perspectives on democracy connected to the participatory dimension are articulated and varied. In this essay, I focus only on those theoretical frames

---

[3] Founded by *ERA-NET Cofund Smart Urban Future Call 2016*, the SoHoLab project involved the Vrije Universiteit Brussel (international coordinator), DAStU—Politecnico of Milan, AHTTEP—AUSSER—École Nationale Supérieure d'Architecture Paris La Villette. Additional nonacademic partners at a local level were included as well. For more information, see the website: www.sohola b.org.

that open up different perspectives and consequent ways of understanding participation as a political context in planning, in which a multiplicity of actors, not only the public institutions, could have a role as promoters and activators.

A first, well-established frame concerns deliberative democracy, which operates on cognitive resources and on the creation of a relational context. In this frame, on one hand, the public actor can recognize citizens' preferences expressed in structured environments; on the other citizens prove their accountability through their point of view and inquiring about public choices. This process is not necessarily an occasion to remove differences and conflicts, but it can create an arena in which disagreements can be expressed in a reflective and informed way (Floridia 2017). Deliberative thinking is mostly associated with the work of Jürgen Habermas (1996),[4] who considers deliberation to be established by institutions, ensuring the equal participation of all. Habermas is aware of the fact that different cultures, world views, and ethics can lead to difficulties in the deliberative process. Despite that, he argues that communicative reason can create a bridge between opposing views and interests.[5]

Starting from 'deliberative polling' and going up to 'citizens juries' or 'citizens assemblies', there are numerous examples of models and methodologies that try to translate the idea of a 'deliberative democracy' into practice (Bobbio 2010). Furthermore, in some cases, this frame has been consolidated in Europe with the establishment of new rules and laws that regulate the relationship between institutions and citizens in a structured form, for example on issues such as large public works.[6]

Employing a 'radical democracy' perspective, Laclau and Mouffe (1985) argue that deliberative democracy risks flattening differences and repressing points of view by building consensus.[7] They primarily suggest that power relations need to be made

[4] Habermas introduced this concept in policy sciences and urban studies in the 1990s, arguing that political problems surrounding the organization of urban life can be resolved through deliberation: people coming together and deliberating on the best possible solution, developing a reflexive critical process of coming to the best solution. Deliberative democracy is now an influential approach to the study of democracy and political behavior. Its key proposition is that, in politics, it is not only power that counts but also good discussions and arguments.

[5] To solve these difficulties, those who practice this kind of participation emphasize that places and instruments of discussion must be structured with higher care taken in the identification of participants, the offering of debates, the conducting of processes, and the organization of the physical space; citizens should be enabled to obtain balanced information, to consult experts with different backgrounds, and to express their opinions.

[6] Different books examine the interplay between the normative and empirical aspects of the deliberative model of democracy, presenting the main normative controversies in the literature on deliberation, including self-interest, civility, and truthfulness and proposing methods by which deliberation can be assumed as well as measured. See, for example, Gastil and Levine (2005), Steiner (2012), Jossey et al.

[7] Laclau and Mouffe (1985) propose to pursue a more radical vision of democracy, focusing on difference, dissent, and antagonism. The challenge is to strengthen democracy around difference and dissent, looking at ways to redefine existing power relations. This reflects a broader turn to thinking of democracy which focuses on redefining 'the political' as a realm in which new identities are formed and new agendas are generated, and through which official procedures and institutions are contested and potentially transformed.

visible and should be renegotiated and changed, setting important points of attention regarding participation.

A second frame is about policymaking and place-making—concepts linked to the definition of policies and place—through interaction and social participation. These are types of local collaborative processes, aimed at treating a complex problem through the realization of a project, the implementation of a policy, or a small concrete improvement. In critical urban practice, making can start from the concrete transformation of objects and spaces that define it and become a tool to examine a situation and trigger future action for change.

Local experiences referring to a 'collaborative construction of the city' are becoming more and more important for territorial development and cohesion, and therefore can no longer be considered 'marginal' experiences. The concepts of 'doing together' and 'activation from below' are emerging as new words for urban planning (Bianchetti 2014; Wachter 2020). In this sense, we can say that the result of participation is not so much a decision or the opening of areas for discussion and debate, as in the deliberative perspective. Instead, it is an outcome primarily concerned with concrete urban issues (policies, scenarios of change, reuse projects, modification programs, and transformations).[8] This activity does not necessarily refer to the design of spaces, but to design processes that start from places in a broad sense (a small garden, a disused building, an urban plan, an event in public spaces, a community center, etc.); it also refers to the possibility of establishing new links with the territory based upon design and action, upon the construction of collective spaces of identity and self-representation.

In other words, a process of co-design can contribute to social mobilization, expanding the range of devices and rituals that residents and their support networks can put into practice. If approached in a situated and collaborative manner, design can support participatory processes in connecting experiences of everyday life with broader visions for more emancipatory urban futures. It also focuses on open-ended scenarios and options for change, rather than settling for the lowest common denominator (De Carli and Frediani 2021).

Finally, the third theoretical frame looks at experiences promoted in a self-organized form by citizens who participate in the transformation of the city through their practices, outside institutional settings. Actors undertake initiatives that concretely push themselves to confront urban issues, testing their ideas and abilities. These practices position participation beyond formal planning systems and highlight everyday life as a key site for urban change. This is a vast field of activities and experiences that sees the widespread involvement and leading role of citizens, individually or in association, as activators of processes of change starting from the urban space.

---

[8] Design via social interaction is an attempt to construct meaning and a common space, starting from the direct sharing of the 'things to do' related to interests and common goods (Laino 2012). Interaction aimed at a physical structure helps participants develop a real argument consisting of small advances in content, which often take place more easily if they are concerned with the dimension of 'making' and the construction of a common space, which is not a deliberative space but a physical one.

The relationship between collaborative practices and the reappropriation or regeneration of urban spaces is crucial and profound.[9] Even if social processes of this kind have always been present in our cities, today it is extra evident, since these practices are gradually expanding and covering a wider range of diverse fields and actors. Often, activation is linked with everyday actions that circumscribe areas of reappropriation, reclamation, and reinterpretation of the local, indicating 'struggles by communities and individuals' (Hou 2010) to find their place and expressions.

These experiences are also important occasions for 'cultivating sociability', experimenting with relationships, and sharing knowledge and alternative models of consumption: in many cases, they are 'social' experiences, that concern the dimensions of appropriation and construction of common paths (Cellamare and Cognetti 2014). They constitute what Ingold (1993) defines as a 'taskscape'; that is, a social landscape characterized by related human activities that grow in it. These practices are not static, but instead must be considered in a continually changing procedural perspective.

The different approaches to participation and participatory planning presented here are not mutually exclusive, but there are important variations in emphasis between them. They question the role of institutions and of civil society and propose different balances between pushes for change (e.g., between top-down and bottom-up, or between 'way of deliberation' and 'way of action').

2.4 Participation as a Process of Enabling? A Possible Role for Living Labs

The three frames outlined above show different perspectives that contribute to reinterpreting participatory planning as a policy tool, i.e., as the bearer of a concrete conception of the relationship between politics and society (Lascoumes and Le Galès 2007). This avoids reducing it to a technique and efficient procedure for 'involving' participants, valid regardless of the contexts in which it is applied.

The three frames are not exclusive, are all partially valid for reinterpreting Living Labs, and should be considered in order to take the social and political context into account. However, it is useful here to place Living Labs in the second frame, emphasizing their nature as a 'design device' which selects actors, interests, opportunities, resources, and problems and helps problematize the nature of the actors involved (public, political, technical, social).

Indeed, Living Labs, as described in Chap. 4 of this book, are a device to reinforce people's ability to reflect on, aspire to, and take action for the transformation of their life context, becoming real agents of change (and in this aspect, they are part of the

[9] Usually, these practices promote the regeneration of empty buildings or abandoned green areas or foster the organization of informal welfare and accessible cultural production.

third frame described above). Living Labs are therefore a space for building networks and alliances for change and for mutual learning and capacitation through action and design.

In this sense, they can contribute to nurturing an enabling process: a long progression and practice of education, a kind of 'collaborative learning arena' that builds links between narratives, competences, organizational logics, and different cultures. This 'gym' is a learning environment linked to the construction of interpretations, the realization of plans or the drawing of scenarios, and the implementation of concrete projects. Mutual contact and familiarity, collaboration on knowledge construction, and concrete actions of change thus become levers for learning from others and from the situation. In this way, the process stimulates improved institutional learning on the one hand. On the other, the learning practice is broader and includes other actors who progressively qualify as relevant partners (third-sector organizations, local groups, residents, etc.). A relevant outcome is also the possibility of scaling up and introducing learning into ordinary practices (both institutional and local).

To initiate this process, the question of the positioning of experts is central: they are called upon to be part of the learning dynamic, making their expertise available, while practising involvement and proximity. The figure of the expert required by this process is therefore not so much that of a facilitator but rather that of an enabler who takes part in the dynamics of the territory, inhabiting the places, within the existing power relations, close to residents and local institutions.

In some experiences, this figure is represented by university bodies, which can play an expert but 'third' role, taking part in processes with a propensity for self-reflection and cultural autonomy. Moreover, this figure is an actor which, by its very nature, has a vocation linked to education and the formation of skills. In this sense, the university can play a crucial role as a knowledge bridge (Benneworth and Cunha 2015): on the one hand, it becomes an enabler, legitimizing local competences and capabilities; on the other, it reinforces its role as a responsible actor among others by working directly with local operators and communities.[10]

Of course, there are many definitions of a Living Lab: it is an 'umbrella' concept that contains many aspects and different approaches, reflecting the same risks to which participation is exposed, as described above. Through the SoHoLab project, developed in contexts that pose specific questions to participation, we have developed a type of intervention that I will illustrate in the next section.

[10] This condition entails finding new ways of interaction between the university and society that can overturn the one-way relationship traditionally conceived: from the idea of the city as an object of study for the university, to the idea of the city as a partner with whom the academy can build a dialogue useful for refreshing academic reflections and enhancing the collective understanding of urban practices. A new current in Italy identifies this as a society-oriented component of the Third Mission, which gives an increasing amount of value to scientific research and innovation as tools capable of supporting society in overcoming the multidimensional challenges it faces (Cognetti 2021).

The definition of the European Network of Living Labs (ENoLL) is similar to many others that can be found in the literature about Living Labs.[11] For instance, Leminen and Westerlund (2012, p. 7) define Living Labs as 'physical regions or virtual realities in which stakeholders form public–private-people partnerships of firms, public agencies, universities, institutes, and users all collaborating for prototyping, validating, and testing of new technologies, services, products, and systems in real-life contexts'. This very broad definition highlights certain aspects in the field of citizen participation, mainly linked to certain characteristics: the idea of generating and nurturing a multifactor ecosystem; the possibility of practising a real-life oriented approach; and the opportunity to initiate a process linked to co-design, which integrates research and innovation.

A first theme is therefore related to the potential of Living Labs as multi-actor platforms, cross-boundary objects/arenas, and contexts for the creative use of knowledge that can connect stakeholders and relevant actors at different levels (institutions and the so-called 'users', for example, citizens and communities). In this perspective, Living Labs are devices shared by various stakeholders[12] to cocreate knowledge for sustainable products and services in real-world settings (Evans et al. 2015). The forms of interaction among the different actors may be different. Leminen et al. (2014) identify four user roles in Living Labs—informant, tester, contributor, and cocreator—in order to underline the possible ways of participating with different degrees of involvement, influencing innovation (Nyström et al. 2014).

A second theme relates to the real-life environment and thus the environment in which the participatory process takes place. Indeed, Living Labs are described as real-life environments in which to experiment, develop, cocreate, validate, and test services and systems with different actors. Environments range from a single isolated place to broader environments such as educational institutes, people's homes and workplaces, and even a city or a part thereof. Real-life environments play a role as landscapes intertwined with stakeholder activities (Leminen and Westerlund 2016), a sort of familiar usage context (Schuurman et al. 2015) in which to experiment with actions.

Finally, Living Labs have a strong action orientation, as they help develop new products and services by engaging users with heterogeneous knowledge, ideas, and experiences. 'User' involvement indicates a shifting of innovation toward the community, thus cocreating with them (Tukiainen et al. 2015). Overall, Living Labs represent

[11] The ENoLL is the international federation of benchmarked Living Labs in Europe and worldwide. Living Labs are defined as user-centered, open innovation ecosystems based on a systematic user co-creation approach that integrates research and innovation processes in real-life communities and settings. In practice, Living Labs place the citizen at the center of innovation and have thus shown the ability to better mold the opportunities offered by new ICT concepts and solutions to the specific needs and aspirations of local contexts, cultures, and creativity potentials. ENoLL identifies five key elements that should be present in a Living Lab: (1) active user involvement; (2) real-life setting; (3) multistakeholder participation; (4) a multimethod approach; and (5) cocreation.

[12] Such as academics, developers, industry representatives, and citizens as well as various public, nonprofit and private organizations (Ballon and Schuurman 2015).

a promising tool to stimulate cocreation by including diverse target groups and often including just a small number of users with specific characteristics.

2.5 Socially Oriented Urban Living Labs. Reflecting on Urban and Social Connotation

We have seen that Living Labs have potential in terms of participation because they emphasize processes of network composition for urban governance, immersion in real life, and co-creation of knowledge and action. But under what conditions can they be enabling spaces? This is a central question in marginalized contexts, which are often the site of failures, mistakes, and interrupted projects. Therefore, more than other contexts, they require support for institutional and social capacities to generate new paths to growth and learning.

The SoHoLab project has created an interesting domain for experimentation, in which different planning and policy tools have been tested through Living Labs operating in fragile contexts, such as social-housing estates. Interpreting the project outcomes, I propose to talk about socially oriented ULLs; that helps to stress the term 'urban' and the labs' social connotation,[13] which generates a process of enabling.

Living Labs are urban because they refer to the political dimension of cities. In this perspective, they are environments allowing those involved to experiment and train the civil capacity 'of all', thus challenging the functioning of local democracy. This implies a focus as much on residents and local players as on institutions. All these actors must be put in a position to learn from the local situation, introducing mechanisms that contribute to the redefinition of both the habits and the places of citizenship as well as routines and institutional norms.

Living Labs also favor the social dimension of planning: they call into question for involving and supporting fragile profiles, which are often excluded from the political process. In marginal contexts, the risk is that the arena has 'selective access', in the sense that only those forces (institutional, private, third sector, associations) that can enter the areas of discussion will participate, without a substantial redefinition of power relations and democratic arrangements. The paradox is that Living Labs themselves can become an instrument of exclusion, particularly of the weakest actors, who do not have the tools to represent themselves.

Therefore, one field of work is the recognition of informal and fragile actors: a work that identifies and favors the emergence of those subjectivities that can bring new knowledge and can acquire new roles in local contexts. There are in fact significant disparities between people, which the process must take into account: they

[13] Elena Maranghi and I introduced aspects relating the prefix 'social' to Urban Living Labs in Chap. 4 in this book, emphasizing the possibility of forming a platform that rehabilitates residents of peripheral contexts first by recognizing them as bearers of desires, capacities, and wills. So what does it mean to place the emphasis on the prefix 'social'? In this regard, the chapter says that 'social' ULLs promote actions that are cultural and plural and that unfold on a daily basis. The definition 'socially oriented Urban Living Lab' was also taken up by Aernouts et al. (2020b).

concern their capacity for self-representation, the quality of the resources they are able to provide, their practical knowledge of the problem to be dealt with, and their ability to formulate their ideas adequately. The habit of participation and the ability to share resources and develop an approach to design thinking does not apply to everyone. Therefore, it is important that the Living Lab is socially oriented and that it takes on the real possibilities of expression and choice developed within the process itself.

So, when can Living Labs be considered socially oriented ULLs? First of all, when they are able to stimulate complex interactions among institutions and local actors belonging to civil society (bringing different pieces of knowledge, values, cultural belongings, power positions, etc.). Indeed, they are primarily intended as a relational field, in which relationships of trust and cooperation can be strengthened and can enrich the local social capital. As such, the formation and nurturing of a collaborative network that promotes visions and policies are crucial, prioritizing the process rather than the product. Too often, in fact, the ability of actors to collaborate and exchange is taken for granted. Latent conflictual dimensions, competitive and power dynamics, antithetical and difficult-to-reconcile positions are often not taken into account. Socially oriented ULLs work on that, favoring the establishment of a collaborative network, a locally rooted 'community of planning' (Maranghi 2019), which is constituted through an investigation activity aimed at planning.

Second, Living Labs can be considered socially oriented ULLs when they are fluid and incremental contexts, in which actors can change their positions, roles, and points of view over time through different kinds of possible interactions (cooperative but also conflictive). Relationality offers a mutual learning path generated through the recognition and acknowledgment of a field in which different interests can be negotiated and conflict can be seen as a manageable issue rather than a barrier. In places such as large-scale social-housing estates—where the traditional relationship between political representation, routine mechanisms of governance, and social cohesion are often compromised—these socially oriented ULLs can contribute to 'mending' relationships. This is particularly relevant in contexts where institutions (at different levels) have progressively lost their ability of 'staying in contact' with the local and thus the effectiveness of their actions.

In this perspective, it is useful to be aware of the risk of referring to the enabling space as a conciliatory place, aimed at pacifying conflicts and building consensus, perhaps more sensitively; traps in which there is a risk of not practicing more radical oppositional positions in order to preserve established relationships and small positions of acquired power. In some cases, on the other hand, empowerment might actually derive from a process that is not collaborative but conflictual, as an opportunity to nurture a widespread critical spirit, to make positions explicit, and argue for different points of view.

Third, Living Labs can be considered socially oriented ULLs when they are contexts able to produce new knowledge because of the construction of bridges among social worlds. This aspect implies that all forms of knowledge are legitimate within the platform. Through participation in these processes, actors are induced to negotiate their own values, roles, and understandings and are, therefore, 'enabled' to build new ones, based on the relationships with other actors involved.

An environment is generated that fosters sensemaking, i.e., a knowledge-building process that constructs new meanings with respect to everyday facts and situations in the lives of people (Weick 1995). This can also take place retrospectively, through a 'back-talk conversation' (Rein and Schon 1994) which, by means of a reflexive confrontation, contributes to reading the past and improving the ability to forecast and develop the future. In fact, the effort is oriented toward understanding the levers of change in a specific area and therefore toward a shared focus on what the problems and possible solutions are.

2.6 Situating and Permanence: Embodied and Rooted Socially Oriented Urban Living Labs

Socially oriented ULLs create new types of collaboration based on local spheres of democracy, a trading zone that acts as an 'exchanger' for dialog among different subcultures (Balducci and Mantysalo 2013; Gorman 2010), building bridges and coherence between different worlds (municipalities, different institutions, associations, citizens' groups, individuals, etc.) and between different scales of planning.

These physical or virtual places have emerged in some recent reflections on participation as 'neighborhood think tanks' (Laws and Forester 2015): on the one hand, they pay attention to structural inequalities and urban-scale dynamics taking shape in the territories; on the other, they take care of local conditions for change, such as the establishment of new coalitions for community development, the strengthening of leadership, and capacity-building. They have a transformative role, in which mutual exchange triggers the possibility for change and improves existing practices.

In the three SoHoLab national contexts, the research units experimented with different forms of 'being in the field', with the common characteristic of somehow leading the researchers to 'inhabit the place' to a certain extent.[14] We can define such an approach as 'situating' (Castelnuovo and Cognetti 2019); in this regard, other authors have referred to the concept of 'architectural permanence' (Hallauer 2015),

[14] In the Belgian case, one of the researchers conducted ethnographic fieldwork between July 2017 and July 2018, during which she lived in the neighborhood, in an empty apartment provided by a social housing company; in the Italian case, the research group reopened a vacant space in the neighborhood in 2014, which is currently open three times a week and which has been operative during the last 8 years. The French team was also 'on-site' to a certain extent, both through the ethnographical survey realized in the La Fosse aux Loup neighborhood and during a three-day-long experimental, interdisciplinary on-site workshop in La Gonflée, which involved architects, town planners, artists, sociologists, etc.

derived from the culture of the 'artist in residence',[15] or 'exploring the embodied side of cities' in the field of cultural and urban studies.[16]

Choosing a situated approach to planning involves centering live experiences, particularly those of exclusion, rather than 'technical attitudes'; approaching sites and people through careful listening and observation; and working with others to understand situations together and from within. The notions of 'situated knowledge' (Haraway 1988) and 'situated learning' (Lave and Wenger 1991) highlights that knowledge always reflects the perspectives of the knower and the relationship between learning and the social situation in which it occurs. Through reconnecting with 'situated stories' (Doucet and Frichot 2018),[17] other forms and imaginations of engagement can emerge, reclaiming a capacity for agency in situations where power relations have become imbalanced.

These concepts suggest that establishing the ULL as a stable presence in the local context is critical for at least two main reasons. On the one hand, in large social-housing estates, a long-term and stable presence counters the processes of abandonment and neglect that characterize these places. Therefore, it helps build a qualified relationship with a context in which residents and local organizations usually perceive research and institutions as temporary and unstable presences, often 'taking' from the context without offering anything in return. On the contrary, taking care of a certain place or 'simply being there' (Aernouts et al. 2020a), listening to people and directly facing what they are experiencing in relation to the place, helps practice a relational dimension based on trust between and mutual recognition of both residents and local organizations.

On the other hand, inhabiting a place can help deal with the 'unexpected in the contingency' (Cognetti 2018) that fosters the collection of the most diverse materials as clues.[18] In other words, 'being on site', in different ways, helps deal with the complexity of social-housing contexts, where the overlapping of different dynamics and phenomena makes it difficult to frame issues and problems clearly. This helps avoid the pitfall where participatory planning tends to merely answer predefined questions from an 'expert' and external perspective. The contingencies associated with being on a site enable unexpected questions to pop up. Contingency (Karvonen and Van Heur 2014) also helps to critically look at the context from

[15] According to which artists reside in a certain place to create pieces on and within it.

[16] In the field of urban studies, there is a growing stream exploring the embodied side of cities (Careri 2002; Degen and Rose 2012; Hubbard and Lyon 2018; Low 2015; Paterson and Glass 2020), which seeks to provide detailed accounts of the ways in which walking and other forms of embodied presence produce understanding and meaningful design.

[17] The authors are coeditors of a special issue of *Architectural Theory Review*, *22*(1), 2018, for which they invited contributors to examine the potential of situated perspectives for the study of architecture and the city and to demonstrate the possibility of critical engagement in research and design through the analysis of concrete practices and practices of thought: architectural and urban, contemporary and historical.

[18] We can also refer here to the concept of 'floating observation' (Petonnet 1982), intended as a methodological tool that allows researchers to approach the site with an open and ever available perspective, enabling information to penetrate without a filter until reference points, convergences, and underlying rules emerge (see Aernouts et al. 2020b).

an 'internal perspective', which is open to new interpretations and meanings and which at least tries to overcome stereotypical representations often associated with large-scale social-housing estates.

In other words, the process and practice of situating help build a more 'reliable' form of knowledge: by changing position and temporarily becoming 'residents', urban experts, researchers, and practitioners change their perspective, diving into the everyday life and daily practices of the context (Cognetti and Padovani 2017). Theoretical concepts, rather than being mobilized as analytical 'lenses' through which to study the world, 'come from specific situations and are transformed by the subsequent situations in which they are deployed, that is to say, with each new situation, a concept is likewise resituated' (Doucet and Frichot 2018, p. 3).

Being part of the place helps researchers develop a vision of their role, which includes loving attachment to people and place. By 'love', here I refer to its multiple dimensions of 'trust, commitment, care, respect, knowledge and responsibility (Porter et al. 2012, p. 603). Moreover, a long-term presence, mixed with a scientific and critical background, can function as an important resource for effective interpretations of problems or creative elaborations of possible answers. In this way, socially oriented ULLs become able to bridge different 'positions' and social worlds (academic, institutional, local, etc.) because they can understand the different sides, while (and precisely because) they do not completely belong to any of them.

These practices are rooted in everyday life; more specifically, social ULLs can be placed in a geographical area—authors refer to geographical embeddedness (Voytenko et al. 2016)—within which the lab represents the ecosystems of open 'urban' and 'civic' innovation that takes place in a real urban context.[19] So, even if it is limited in time or objective, the dimension of 'diving into' a real-life context is considered extremely relevant.

The focus then is on how this 'immersion' takes place. Social ULLs should be framed as 'spaces of encounter', sufficiently open to unexpected interaction and to the unplanned. They should be experimental environments, not entirely artificially constructed with a top-down (even if participatory) approach but emerging from the encounter between researchers and users. Their overall outcome is not predefined but is the result of the interactions between promoters and stakeholders.[20]

According to Franz (2015), time is a crucial dimension in the promotion of effective local involvement.[21] The author questions the feasibility of the limited duration

[19] This may be a region, an agglomeration, a city, a district or neighbourhood, a road or corridor, or a building. There are many possible urban configurations that can host a ULL, but the area is normally clearly defined and has a manageable scale (Voytenko et al. 2016).

[20] Also, in the space of the ULL—whether it is a stable and physical space or a more metaphorical space of encounter—a balance should be constantly pursued between a planned space and a space that is flexible enough to adapt to what comes from the context in terms of inspiration, demands, and suggestions.

[21] 'Social living labs should ensure authenticity and credibility. Both cannot be assured as long as the research is limited to the duration of a specific research project. To create a trusting and collaborative interaction with local citizens, a shift in research strategy towards long-term engagement is unavoidable' (Franz 2015, p. 113).

of her research and the outcomes that it generates in the local context. The participatory process has to incorporate 'timing issues', also promoting initiatives taken to activate other organizations such as local housing associations, the regional housing corporate, the office of the government architect, the municipality, and local social workers in order to continue initiatives on a more permanent basis. Therefore, we could state that if the conditions do not enable a long-term presence, other devices must be found to assure a certain continuity of engagement or to engage other actors to take up this role.

2.7 Conclusions

Socially oriented ULLs are local planning platforms in which it is possible to experiment not only with good practices among citizens and local organizations but also with new forms of governance and interaction among administrations. In this way, socially oriented ULLs can help assess the validity of those processes at the local level, where institutions are pushed to experience the possibility of innovation of procedures, and locals have the chance to strengthen their community empowerment. In this sense, the context enables people to 'do' and build visions for the future. By strengthening the networks and capacities of actors, areas with ULLs can be considered experimental laboratories for the future: the future, in fact, becomes a 'cultural fact' (Appadurai 2014), contrasting the 'ethics of probability' with that of 'possibility'.

Initiating complex processes requires new intelligence, skills, and sensitivity from everyone: a new openness and attention to the creation of multilevel and multi-actor planning areas and tools from public institutions; an open and inclusive idea of design in which the city becomes interlocutor for a collective enterprise that develops over time from the experts; an active and proactive role beyond the historical inertia and opposition from the position of residents and local forces; sensitivity and attention to important opportunities to be seized from the local players. This is a very ambitious horizon, because it is linked to the possibility that during the process, spaces will open up both for learning (individual and collective) and for generating new protagonists (individual and collective).

In conclusion, in contrast to the more consolidated forms of participation which, as mentioned before, trigger deliberative processes promoted unilaterally by institutions supported by facilitators, the paper introduces two other ways of understanding participation. These two ways are promoted not only by public actors but also by many others who participate in the transformation of the city through their practices. These modes are based on co-design and collaboration between different partners or promoted in a self-organized form by citizens, outside institutional settings.

Living Labs have the required characteristics to be part of the field of co-design and reciprocity between different actors, with a focus on the protagonists as individuals and groups. For this reason, they are positioned between the second and third modes, emphasizing their nature as a 'design device', reinforcing people's ability, and building networks and alliances for change and mutual learning.

In this arena, experts are also called on to take on new roles, taking care of local conditions for changes such as the capacities of individuals to become active, the establishment of new coalitions for community development, and the institution building.

If Living Labs promote processes of interaction between different actors oriented toward co-design, the urban dimension, and the social connotation are central. Therefore, the paper proposes to call them socially oriented Living Labs. This is necessary in marginalized contexts, in which it is useful to work on a social context that encourages people and their capacities, supporting the building of agency in situations where power relations and expectations have become imbalanced.

It seems necessary to start reflections and experiments in this direction in European urban contexts increasingly characterized by social and economic inequalities and by marginal contexts that require participatory planning to be radically rethought, addressing both theory and practice.

References

Aernouts N, Maranghi E, Ryckewaert M (Eds) (2020a) The regeneration of large-scale Social Housing estates. Spatial, territorial, institutional and planning dimensions. SoHoLab

Aernouts N, Maranghi E, Ryckewaert M (Eds) (2020b) Towards a definition of socially oriented LivingLabs. SoHoLab

Appadurai A (2014) Il futuro come fatto culturale. Saggi sulla condizione globale. Raffaello Cortina

Balducci A, Mantysalo R (Eds) (2013) Urban planning as a trading zone. Springer

Ballon P, Schuurman D (2015) Living labs: concepts, tools and cases. Info 17(4):1–11

Beebeejaun Y (2006) The participation trap: The limitations of participation for ethnic and racial groups. Int Plan Stud 11(1):3–18

Benneworth PS, Cunha J (2015) Universities' contributions to social innovation: Reflections in theory and practice. Eur J Innov Manag 18(4):50–52

Bianchetti C (2014) Territori della condivisione. Una nuova città. Quodlibet

Bobbio L (2010) Types of deliberation. J Public Deliberation 6(2), art. 1. https://doi.org/10.16997/jdd.105

Bobbio L (2016) La partecipazione imperfetta. Sentieri Urbani, Rivista Della Sezione Trentina Dell'istituto Nazionale Di Urbanistica 2(1):30–33

Brownill S, Parker G (2010) Why bother with good works? The relevance of public participation(s) in planning in a post-collaborative era. Plan Pract Res 25(3):275–282

Careri F (2002) Walkscapes. Editorial Gustavo Gili, El andar como pràctica estética / Walking as an aesthetic practice

Castelnuovo I, Cognetti F (2019) Mapping San Siro Lab: experimenting grounded, interactive and mutual learning for inclusive cities. Trans Assoc Europ Schools Plann 3(1):37–54

Cellamare C, Cognetti F (Eds) (2014) Practices of reappropriation in the contemporary city. Processes, places and imaginary. Tracce Urbane, Planum Publisher

Cognetti F (2018) Apprendere la città. Le prospettive di una ricerca inclusiva e relazionale. In Cognetti F, Padovani L (Eds), Perché (ancora) i quartieri pubblici. Un laboratorio di politiche per la casa. Franco Angeli, pp 33–51

Cognetti F (2021) Responsible research and responsible education: Society-oriented components of the third mission. In Orsini F, Pogliani L (Eds), The future of peripheries. Re-shaping the metropolitan city. Pearson Italia, pp 52–57
Cognetti F, Padovani L (2017) New meanings for public housing through the co-production of knowledge. Policies for everyday life in marginal neighbourhoods. In Bargelli E, Heitkamp T (Eds), New developments in Southern European housing. Pisa University Press, pp 53–77
Concilio G (2016) Urban living labs: opportunities in and for planning. In: Concilio G, Rizzo F (eds) Rethinking the interplay between design and planning. Springer, pp 21–40
Cooke B, Kothari U (2001) Participation: the new tyranny? Zed Books
Crosta PL (2003) A proposito di approccio strategico. La partecipazione come tecnica di pianificazione o come politica di cittadinanza attiva? In Moccia FD, De Leo D (Eds), I nuovi soggetti della pianificazione. Franco Angeli, pp 239–245
De Carli B, Frediani AA (2021) Situated perspectives on the city: a reflection on scaling participation through design. Environ Urban 33(2):376–395. https://doi.org/10.1177/09562478211028066
Degen M, Rose G (2012) The sensory experiencing of urban design: the role of walking and perceptual memory. Urban Studies 49(15):3271–3287
Doucet I, Frichot H (2018) Resist, Reclaim, Speculate: situated perspectives on architecture and the city. Architect Theory Rev 22(1):1–8
Evans J, Jones R, Karvonen A, Millard L, Wendler J (2015) Living labs and co-production: university campuses as platforms for sustainability science. Curr Opinion Environ Sustain 16:1–6. https://doi.org/10.1016/j.cosust.2015.06.005
Floridia A (2017) From participation to deliberation. ECPR Press, A critical genealogy of deliberative democracy
Franz Y (2015) Designing social living labs in urban research. Info 17(4):53–66
Fung A (2017) Putting the public back into governance: The challenges of citizen participation and its future. Public Adm Rev 75(4):513–522
Gastil J, Levine P (2005) The deliberative democracy handbook. Strategies for effective civic engagement in the twenty-first century. Jossey-Bass
Gorman M (Ed) (2010) Trading zones and interactional expertise: creating new kinds of collaboration. MIT Press
Habermas J (1996) Between facts and norms. MIT Press, Contributions to a discourse theory of law and democracy
Hallauer E (2015) Habiter en construisant, construire en habitant: La 'permanence architecturale', outil de développement urbain? Métropoles, 17. https://doi.org/10.4000/metropoles.5185
Haraway D (1988) Situated knowledges: The science question in feminism and the privilege of partial perspective. Fem Stud 14(3):575–599
Hou J (2010) Insurgent public space: Guerrilla urbanism and the remaking of contemporary cities. Routledge
Hubbard P, Lyon D (2018) Introduction: Streetlife—the shifting sociologies of the street. Sociol Rev 66(5):937–951
Ingold T (1993) The temporality of the landscape. World Archaeol 25(2):152–174
Karvonen A, Van Heur B (2014) Urban Laboratories. Experiments in reworking cities. Int J Urban Regional Res 38(2):379–392
Laclau E, Mouffe C (1985) Hegemony and socialist strategy: towards a radical democratic politics. Verso
Laino G (2012) Il fuoco nel cuore e il diavolo in corpo. La partecipazione come attivazione sociale. Franco Angeli
Lascoumes P, Le Galès P (2007) Introduction: understanding public policy through its instruments. From the nature of instruments to the sociology of public policy instrumentation. Govern Int J Policy Administr Instit 20(1):1–21
Lave J, Wenger E (1991) Situated learning. Cambridge University Press, Legitimate peripheral participation

Laws D, Forester J (2015) Conflict, improvisation, governance: street level practices for urban democracy. Routledge
Lefrançois D (2021) From participation to livinglab. A look back at experimental approaches to refurbishment, multi-partner projects and participatory initiatives. SoHoLab
Legacy C (2017) Is there a crisis of participatory planning? Plan Theory 16(4):425–442
Leminen S, Westerlund M (2012) Towards innovation in Living Labs networks. Int J Prod Dev 17(1–2):43–59
Leminen S, Westerlund M (2016) A framework for understanding the different research avenues of living labs. Int J Technol Mark 11(4):399–420
Leminen S, Westerlund M, Nyström AG (2014) On becoming creative consumers–user role in living labs networks. Int J Technol Mark 9(1):33–52
Low KEY (2015) The sensuous city: sensory methodologies in urban ethnographic research. Ethnography 16(3):295–312
Maranghi E (2019) Networking collective knowledge to foster change. The case of the Sansheroes network (San Siro, Milan). In: Proceedings of the annual conference of the associations of european schools of planning—planning for transition, 13 July 2019, Venice, AESOP
Maranghi E, Cognetti F, Ryckewaert M, Aernouts N, Mosseray J, Lefrançois D, Wachter S, Labied N (2020) 5 themes from urban living lab experiences in large-scale social housing estates. SoHoLab
Mela A (2016) Partecipare alle trasformazioni urbane in un'epoca di transizione. Sentieri Urbani 8(21):18–21
Miessen M (2010) The nightmare of participation. Sternberg Press, Crossbench praxis as a mode of criticality
Nadin V, Stead D, Dąbrowski M, Fernandez AM (2021) Integrated, adaptive and participatory spatial planning: trends across Europe. Reg Stud 55(5):791–800
Nyström AG, Leminen S, Westerlund M, Kortelainen M (2014) Actor roles and role patterns influencing innovation in living labs. Ind Mark Manage 43(3):483–495
Paterson M, Glass MR (2020) Seeing, feeling, and showing 'bodies-in-place': exploring reflexivity and the multisensory body through videography. Soc Cult Geogr 21(1):1–24
Petonnet C (1982) l'Observation flottante. L'exemple D'un Cimetière Parisien. L'homme 22(4):37–47
Porter L, Sandercock L, Umemoto K, Bates LK, Zapata MA, Kondo MC, Zitcer A, Lake RW, Fonza A, Sletto B, Erfan A (2012) What's love got to do with it? Illuminations on loving attachment in planning. Plan Theory Pract 13(4):593–627
Rein M, Schon D (1994) Frame reflection. Basic Books
Schuurman D, De Marez L, Ballon P (2015) Living Labs: a systematic literature review. Open Living Lab Days
Steiner J (2012) The foundations of deliberative democracy. Cambridge University Press, Empirical research and normative implications
Tukiainen T, Leminen S, Westerlund M (2015) Cities as collaborative innovation platforms. Technol Innov Manag Rev 5(10):16–23
Voytenko Y, McCormick K, Evans J, Schwila G (2016) Urban Living Labs for sustainability and low carbon cities in Europe: Towards a research agenda. J Clean Prod 123(1):45–54
Wachter S (2020) Living Labs, a participatory turning point for landscaping? In: Aernouts N, Maranghi E, Ryckewaert M (eds) Towards a definition of socially oriented LivingLabs. SoHoLab, pp 19–30
Weick KE (1995) Sensemaking in organizations. Sage Publications

Chapter 3
Governing with Urban Living Labs

Serge Wachter

Abstract Within the European Union, Urban Living Labs now figure prominently in the urban governance toolbox. On the local stage, they are seen as auxiliaries that encourage citizen participation with a view to co-constructing policies designed to improve the quality of housing and the built environment. Their versatility is remarkable, and they are used both to boost new approaches to planning, such as tactical urbanism, and to renew the regeneration policies of large social housing areas. This text is the result of the SoHoLab research project and aims to provide a critical analysis of the roles and functions of Urban Living Labs as support tools for planning policies. Based on a review of the specialized literature and the work and experiments carried out within the framework of SoHoLab, it proposes an approach that strives to analyse ULLs as components of a new model of urban governance. From a critical perspective, it formulates the hypothesis that ULLs are not only tools of power aimed at promoting the empowerment of residents but also seeking to improve the legitimacy of planning policies and to impose a model of domination and forms of social control in accordance with the requirements of neoliberal city regulation.

Keywords Governmentality · Urban governance · Neoliberal city

3.1 On the Way to Experimentation and Innovation

The emergence of ULLs as constituent parts of a new model of governance has been stimulated by the combination of a series of factors, creating an 'ecosystem' that has allowed them to flourish. In particular, the uncertainties and tensions linked to the risks generated by the ecological and climate-change-related transitions have prepared a solid ground for reorienting planning practices. This vision of risk has also opened up a space of opportunity by creating a new urban offering that ULLs have been able to infiltrate.

S. Wachter (✉)
École Nationale Superieure d'Architecture de Paris-la-Villette, Paris, France
e-mail: serge.wachter@club-internet.fr

N. Aernouts et al. (eds.), *Urban Living Lab for Local Regeneration*,
The Urban Book Series, https://doi.org/10.1007/978-3-031-19748-2_3

In this respect, Harriet Bulkeley believes that experimental methods in the field of local policy, in particular, those aimed at fighting climate change, are booming (Bulkeley et al. 2011). They especially target planning policies and lend themselves to approaches that put hypotheses to the test in a concrete sense. They provide the means to apply a trial-and-error method and to adjust tools through successive approximations. In so doing, these approaches follow the principles of incrementalism and of the continuous adaptation of intervention modes. Such flexibility is a key asset or resource that facilitates the adjustment and management of development projects. It provides room for manoeuvre when dealing with the dilemmas, conflicts and contradictions found in housing and urban policies (Severini et al. 2015). In short, and to echo four experts, we must recognize that ecological and energy transition policies include ingredients that essentially revolve around the following key words: 'experimentation, learning, reflexivity and reversibility' (Von Wirth et al. 2019). These characteristics form the pillars of a new method for regulating local public action.

From this standpoint, it is important to emphasize that both theoretical approaches and empirical studies carried out in different urban and organizational contexts show that Living Labs, or more precisely Urban Living Labs (ULLs), are highly experimental tools. Their hybrid nature and their position on the fringes or at the intersection of various local public action 'ecosystems' make it possible for them to encourage new institutional arrangements and innovative cooperation. This naturally supposes specific properties of conception or design and their operational implementation. Observing a cycle of experiments related to an Urban Living Lab (ULL) set up in Greater Manchester, specialists carrying out 'action research' concluded that this example of experimentation illustrated 'how the practice of ULLs may involve different forms and techniques for learning, shielding, nurturing, empowering and participating within ULL' (Bulkeley et al. 2011). Moreover, drawing on the lessons learned from the implementation of 40 ULLs in Europe, influential and no less eminent experts in the academic arena declared ex abrupto that ULLs are excellent tools with which to steer experimental towns. It is important to note that the term 'experimental city' is new and revealing. I believe it is rich in meaning, as it underlines the growing importance of experimentation in the 'software' of local urban planning policies. Above all else, it introduces—dare I say it—an epistemological watershed that invites us to observe and analyse interventions in the field of planning from a fresh perspective. In the light of the new approaches shown in these case studies, it must therefore be admitted that urban policies deserve to be increasingly considered and studied—and implemented—through the prism of experimentation. Indeed, 'the recent upsurge of interest in the experimental city as an arena within and through which urban sustainability is governed, marks not only the emergence of the proliferation of forms of experimentation—from novel governance arrangements to demonstration projects, transition management processes to grassroots innovations—but also an increasing sensibility amongst the research community that urban interventions can be regarded in experimental terms' (Marvin et al. 2011).

In this light, authors who are familiar with the development of recommendations and roadmaps for ULLs Labs clearly state that ULLs are considered spaces to facilitate experimentation about sustainability solutions' (Von Wirth et al. 2019). Has experimentation become the 'ultima ratio' of local policies and of planning policies in particular? There is no doubt that, according to this vision, ULLs stand out, if not as the key component of these experimental approaches, then at least as precious auxiliaries seeking to integrate citizens' voices into ecological transition programmes. It is undeniable that a little local democracy cannot harm planning programmes. In reality, experimentation is characterized not only by practical activities but also by a cognitive process that leads to the production of shared knowledge. The result is a common wealth of data, reflections, and knowledge through which a dialogue can be initiated around a 'shared' project. Experimentation offers support and a vehicle for the consideration of long-term visions and short-term measures in the form of concrete actions and practices. It relates to collective research and exploration, a maieutic in which 'a broad suite of stakeholders like firms, universities and actors from government and civil society are navigating, negotiating (and ideally) reducing uncertainty about new socio-technical innovations through real-world experiments, gaining knowledge and experience along the way in an iterative learning-by-doing and doing-by-learning process' (Scholl and Kemp 2016, p. 231).

As we can see, experimentation represents both a conceptual turning point and a reference to empirical practices designed to offer shared operational solutions to the challenges of urban ecological transition. This makes it possible to introduce parameters into these new methodologies that take into account the needs and practices of inhabitants during collaborative planning sessions. These approaches are not limited to the energy transition currently in vogue. By extension, they are spreading to all sectors of local public action and constitute a point of reference, a corpus of best practices that also inspire urban regeneration programmes. It should be noted that the latter are active in places and neighbourhoods where citizen participation is, if not very weak, at least leaving a lot to be desired.

Looking more closely, we should emphasize that ULLs are 'political beings' that are constantly undergoing evolution and transformation. In a way, they are going through a Darwinian selection process, the outcome of which might allow only the best to survive. In particular, they are continually changing the boundaries of their jurisdiction by integrating new partnerships and new actors with whom to explore new modes of action. This experimental nature leads to solutions and scenarios being tested by diversifying local collaborations within new partner networks. This makes it possible to find new ways to test hybrid programmes for urban sustainability. The result is a governance 'puzzle', the components and operation of which are evolving and appear fragile, yet often prove to be surprisingly effective, resilient, and long-lived. This is nothing new. But there is more: these political regimes, as Clarence Stone would say, of which ULLs are a part offer the wherewithal to overcome the classic opposition between top-down and bottom-up approaches to urban planning. Indeed, ULLs' 'intelligent engineering' offers the means to develop a new type of 'lateral partnership' which reinvents collaborative planning approaches. This opens the way to new intersectoral cooperation that creates shared knowledge within a network of actors engaged in local climate policies.

In this respect, it should be noted that experimentation is distinguished as a new mode of governance. It encourages the use of innovative approaches that break away from the routines of 'standard' planning policies that are implemented without risk and which lead to 'guaranteed and secure' results. In this sense, experimental policies turn their backs on the modernist dream or illusion of total control of urban dynamics, which was representative of the golden era of technocratic and centralized planning. It should be noted that these policies combine, into an innovative duo, experimentation and citizen participation in the principles and modus operandi of collaborative planning. De facto, researchers working with Living Labs underline that these experimental practices open the way to new urban regulations that deserve to be analysed in the light of the concept of 'soft governance' or 'informal governance' (Scholl et al. 2016). These expressions are by no means insignificant; they relate to new practices for the coordination of collective action in the field of planning policy. Widespread urban experimentation is thus associated with a new, highly contemporary, and 'very gentle' art of governing people and things. With this orientation, researchers understand ULLs 'as informal and temporary soft spaces of urban governance' (Smas et al. 2016). The latter make it possible to explore various scenarios and opportunities for the application and implementation of soft governance.

It is worth noting that this new mode of regulation is constantly being challenged, revisited, and evaluated in arenas of deliberation that bring together the various protagonists of a 'local political regime'. At the same time, searching for legitimacy, it explores all possible ways and platforms for citizen participation in local decision-making. This is undoubtedly both advantageous and a break away from top-down and 'legal-bureaucratic' visions, to use Max Weber's expression. In particular, City Labs turn their backs on these hierarchical visions. They aim to represent cutting-edge soft governance tools for experimenting with new approaches and practices in the field of 'collaborative planning'. Is this soft governance the expression of a new form of local power which also claims to co-construct the city with its inhabitants? Does it constitute a welcoming ground, a favourable ecosystem for the development of ULLs?

3.2 Living Labs and Tactical Urbanism

Spontaneously or as a result of interests or strategies, new approaches to planning are emerging and adapting to this new regime of budgetary austerity and managerial adjustment orchestrated by the state. The latter is exploring or, more accurately, experimenting with the ways and means of a form of soft governance that aims to incorporate inhabitants opinions into its guiding principles and modus operandi. This trend is of course having a knock-on effect on ways of managing cities and other local public administrations. This is particularly true of the 'tactical' urban planning approaches that have been gaining ground in recent years. Soft governance, the rise of experimental practices, and the development of ULLs are, if not the pillars, then at least the backdrop to these new approaches.

The term tactical urbanism covers a variety of practices that share a collaborative base that generally brings together a range of actors with a view to developing alternative planning practices in order to offer goods and services outside the rationale of market rules. The most prominent of these actors are associations, architects' collectives, and other representatives of civil society. There is a connection between these approaches and the various forms of production and management of common goods that are governed by a collaborative spirit and a direction that is guided more by an ethic of sharing than by the individual appropriation of resources. It is also important to note that temporary urbanism is characterized by the new activities and functions that are put in place in the spaces and buildings they occupy or in which they are involved, identified by the term 'third places'. Living Labs, Fab Labs, and co-working spaces are very often included in the programming and realization of temporary urbanism operations, and it is very common for these activities and services specific to 'third places' to be present on those premises and sites where these ephemeral operations are carried out. It is worth noting that some authors liken these various forms of temporary occupation to a specific category of Living Labs that can be set up more specifically in wastelands or disused spaces in order to develop them and 'co-design' goods and services intended to meet local demand.

These transitional urbanism practices have gained impetus over recent years. Some believe that they open up new avenues for planning policies in which different notions of civic sense and citizenship can find means and channels to express themselves. They emerge and fit into the nooks and crannies of an urban planning and project-based urbanism that is considered to be lacking in innovation and in search of renewal or, even better, 'reinvention'. These 'non-standard' approaches are 'genetically' related to Living Labs and in most cases incorporate their main ingredients: innovation, experimentation, learning-by-doing, and participation of users and the various stakeholders. This trend is nurtured by approaches that promote urban sustainability in all its forms, ranging from shared gardens to short circuits and edible landscapes provided by urban agriculture, along with provisional or temporary installations and architecture in public spaces designed to house and accommodate various collaborative activities that generally follow the 'DIY' and 'grassroots initiative' philosophy. It surfs the fashionable and trendy approaches that advocate recycling and short circuits by blithely denouncing waste, the 'socially irresponsible carbon footprint' of current development and construction activities, and the absence of other efforts to protect the planet. In the same spirit, it promotes the values and practices of collaboration in the fields of uses and services that range from carpooling and the pooling of resources to the sharing of local goods previously managed by the sacrosanct right of ownership. Its members or sympathizers are generally sensitive to fashionable representations advocating a drive towards a 'frugal city' model—not to say a 'slow city'—applying at all levels, for individuals and local communities, the values and practices of sobriety, if not austerity. It should be added that in the rhetoric of the presentation and justifications of these experiments, one often finds a moralistic slant that stigmatizes the 'climatically incorrect' and environmentally insensitive approaches of 'standard' planning visions. These practices of the collaborative economy are of course widely encouraged and stimulated by digital platforms and social networks giving access to information on opportunities for exchange, sharing, and access to local collective goods and services. This collaborative urbanism or even common goods urbanism, which, let us not forget, more often than not operates on the fringes of the dominant urban planning logics, was described in his time by Henri Lefebvre as the expression of the urbanism of men of good will.

Such visions are encouraged by urban coalitions which, willingly or not, stimulate and accompany these alternative practices that break away from 'standard' approaches to planning. These 'urbanism margins' are often incorporated into innovative approaches launched by towns wishing to offer real estate products and hybrid services that unite private developers and collectives or associations. If they are not from the 'social and solidarity sphere', then they at least distance themselves from profit and market rationales. I should add that they embody the practical aspect of an approach and methodology inspired by innovative experimental approaches that are user-centred and participatory and that are eligible to receive the Living Lab label (Canapero and Benavero 2016).

Some observers have interpreted this trend as being a gradual shift from one-off or isolated approaches towards the consecration of a future 'mainstream' of urban policy (Douay and Prévot 2016). This would seem excessive, even though many local authorities have incorporated this type of urbanism into their planning and political agendas. Are these emerging practices likely to take root in the metropolitan landscape? It is true that the 'collaborative urban factory' has found success with indulgent audiences. 'Urban modernists' have indeed become infatuated with this type of experience, which combines informality, creativity, and the festive culture that these approaches convey. In Western metropolises, in particular, this preference expresses 'postmaterialistic' aspirations held by social circles and urban places going through a process of gentrification. It is clear that some local authorities are surfing this wave by capturing and winning the loyalty of sympathizers, followers, or clients. Are we seeing the rise of a new model destined to become, if not dominant, then at least part of common, routine practices that are well placed on the planning agenda? Yes, certainly, but at the scale of micro-projects located on the fringes of an urban production that is, while ordinary, nevertheless the most massively structuring of housing, equipment, and infrastructure operations. For this is the destiny of this alternative urbanism: to intervene on the fringes, in the loopholes and interstices of development policies. Geographically speaking, interstice is the right word, because the elective sites of tactical urbanism are essentially abandoned zones, wastelands, and unoccupied land waiting to be used. Alternative urbanism is popular and high profile, but nevertheless only very moderately affects the essential internal workings of the production and functioning of urban fabrics.

In fact, the visibility and media coverage of these operations are inversely proportional to their real capacity to have significant effects on the urban layout. These projects also struggle to stand the test of time: their nature obeys the temporal cycles of the event and of temporary installations. Either way, this is neither dramatic nor deplorable, as the intent or ambition of temporary urbanism is not to change the architecture of the city. It is important to note that, in the most favourable contexts, these approaches convey values and 'positive feelings' which will spread through imitation and which open up promising avenues and new principles for planning policies. Last but not least, temporary urban planning approaches rely on collective intelligence, the production of common goods, and cooperation; for better or for worse, they disseminate values of sharing and solidarity. In other words, their message is benevolent and offers a glimpse of avenues and possibilities for the creation of planning projects away from the hazards of real estate development. In this respect, they should be congratulated; they deserve to be encouraged, evaluated, and channelled. The rising number of such experiments nevertheless raises one question: is this collaborative model not the prerogative of metropolitanism—even more so in gentrified neighbourhoods with no real power to disseminate into disadvantaged areas, be they social housing neighbourhoods or low-density rural areas? In other words, just like temporary urban planning, Living Labs are still far from having won over the sensitive areas prioritized by city policy (Roux and Marron 2017).

For all of the above-mentioned reasons, at a minimal—one might say cosmetic—level and in many respects, ephemeral urbanism is on the road to institutionalization. Indeed, we can see that many institutions in charge of planning have appropriated the language and codes of this transitory urbanism. The latter now seems to have been incorporated or even digested into/by urban strategies for the production of the so-called 'creative' neoliberal city. All in all, one has to admit that 'tactical urbanism is an essential new element of contemporary urban policy in metropolises that are caught up in the logic of competition to attract investment, creators and tourists' (Douay and Prévot 2016, p. 18).

3.3 Governing with Urban Living Labs

When broadening the picture, one is forced to admit that the political and institutional forces behind this new urban offer—and the instruments that accompany it, such as ULLs—are subject to a new system of relationships between the state and local actors. This model is also influencing social behaviour and the values that drive it by amplifying and intensifying the individualization of society. In France, since the introduction of decentralization, many functions and responsibilities have been transferred to local authorities. This slimming down process has accelerated over recent years, and the state has continued and extended this transfer of important areas of public affairs management outside its sovereign perimeter. As Hélène Reigner points out, the transfer of full responsibility for the implementation and coherence of a growing number of state programmes to local authorities and their territorial partners was achieved in return for new forms of framing and controlling their actions through financial supervision, competition for access to national resources, the setting of standards and the awarding of labels, and the promotion of good practice (Reigner 2013). De facto, such changes are implemented by using new instruments of state spatial intervention that are less prescriptive and that open up a range of choices and opportunities for innovation to local actors: the definition of reference systems and guidelines, benchmarking, calls for projects, and, last but not least, recourse to mediation and participation mechanisms with inhabitants/users, such as Living Labs or ULLs.

Such developments have as their backdrop the rise of remote government, a new form of regulation granting additional management capabilities and initiatives to the private sector and local authorities. The state is withdrawing from its territories, outsourcing functions by redistributing tasks and responsibilities and regulating local public action from afar (Epstein 2006). At the same time, this trend goes hand in hand with the consolidation and weight of a moral and cultural climate conducive to the dissemination of values and norms that accentuate and accelerate the transformation of the social realm into a society of individuals. These trends amplify and strengthen the entrepreneurial watershed that has been at work in Western urban societies since the 1990s.

By extending Epstein's argument, it is possible to view this process as a form of vigorous empowerment of local and urban political systems, a hypermodern modus operandi of power and of technologies of governmentality. To simplify, we can say that in this schema, local actors and more broadly individuals and civil society are invited—obliged?—to show initiative in finding solutions to the problems and difficulties they face. In truth, empowerment—and one of its vectors and supports objectified by the Living Labs—is not only citizen mobilization but also a form of power that releases energies, innovation, and capacities for action. We can see that when they exist or are available, it is a question of bringing out, exploiting, and valorizing local skills and resources. But we need to understand that this process simultaneously leads to domination, coercion, and submission to a new standard, thus placing individuals and the groups they form under a duty to take initiative, be autonomous, and take charge of improving their quality of life. Through the Living Labs, the new spirit of capitalism also infuses local public management and approaches to planning (Boltanski and Chiapello 1999). In the final analysis, do these empowerment practices not reflect the fundamental values and principles of liberal regimes founded on the love of human and citizen rights, with the ultimate aim of achieving freedom and individual emancipation?

It is worth noting that government at a distance is coupled with government by instruments. The latter is illustrated by the implementation of policies for activation and the allocation of duties and responsibilities to inhabitants and users and to their groups on the one hand, and to local actors on the other, cities in particular. In this context, the appearance and rise of Living Labs in urban regulation is one of the instrumental facets, and an illustration, of the rise of this form of government.

But there is more: according to several researchers who have theoretically and empirically investigated the *raison d'être* and modus operandi of many ULLs in Europe in different urban contexts, ULLs can be considered supports for the two main strengths on which urban governance is based. They are the expression of a new way of solving urban problems, one that calls for innovation, experimentation, and citizen participation in situations marked by uncertainty or contingencies and where a multitude of actors are led to deliberate and negotiate in order to reach compromises. This is now well known, and we can agree that ULLs have emerged as possible means of providing solutions to or desirable improvements of situations that pose problems and call for new tools or new ways of trying to solve them. ULLs are thus functional auxiliaries at the service of urban governance and are particularly called upon to solve problems of a practical nature. As Karl Polanyi would say, they are 'embedded' in power relationships and games and are subject to political strategies. This is not really new and follows a pragmatist approach to public action in the tradition of John Dewey. It is true that Living Labs produce shared and reflexive knowledge to nourish academic research, but they are first and foremost tools that aim to offer concrete solutions to problems or contradictions perceived and experienced by actors or stakeholders on a local or urban stage. In this respect, and as we have seen, Living Labs or ULLs stand out as instruments that participate in the implementation of 'soft governance' that favours the regulation of conflictual and cooperative relationships between local actors.

Yet in addition to this, there is a second main thrust that ULLs follow. Their instrumental role goes hand in hand with a moral and symbolic component designed to influence and shape the behaviour of social agents. De facto, this reflects ambitions to improve the 'urban condition' of inhabitants through the use of appropriate socio-economic manipulations that aim to bring about new social behaviour. This second purpose, highlighted by researchers whose critical approach is to be commended, is not trivial (Bulkeley et al. 2019). It is in line with the idea that ULLs are also instruments of power which aim to spread values and a habitus that encourage inhabitants and users to, if not commit themselves, at least accept and comply with the social processes of transitions or changes to achieve greater urban sustainability. In fact, ULLs are the vectors and disseminators of ideas and representations that conform to the ecological correctness of precautionary or even 'frugal' behaviour with regard to the environment. In this respect, a Marxist movement resurrected from the limbo of the past might describe them as 'ideological apparatuses' at the service of a power that aims to coerce, subjugate, and educate. Of course, it goes without saying that ULLs are not disciplinary institutions operating under the influence of injunction or command. Their modus operandi is more incentivizing than coercive. In line with more recent work, such normalization has been analysed as the rise of a neohygienic and neoliberal governmentalization of behaviour (Reigner 2015).

A two-pronged movement is at the origin of this new form of domination. First of all, 'sustainable and inclusive urban development' appears as a post-Fordist avatar of environmental protection at the service of an overarching agenda: that of ensuring an urban climate for business and commerce. This regime prioritizes proactive supply policies to attract postindustrial economy capital and strategic metropolitan jobs. The regime tolerates or even encourages innovative approaches located on the fringes of mainstream planning practices, such as tactical urbanism. In so doing, it uses tools designed to reveal the preferences of residents with the intention of involving them in the design of programmes for the rehabilitation of their housing or their built environment. These programmes happily combine urban sustainability objectives with those of social compensation by mixing the two in a confusing fashion. They straddle social justice and environmental justice. Along the same lines and even more radically, updating the Marxist critique inspired by the fetishism of the commodity, Eric Swyngedouw asserts that 'sustainable urban development is the new opium of the people, a democratic anaesthetic that locks up the spaces of dispute for the expression of contradictions and oppositions' (Swyngedouw 2011, p. 130). This may seem somewhat overstated, but it contains an element of truth. There is no question that the mainstream rhetoric on the frugal, inclusive, and resilient city plays a powerful role in depoliticizing the issues at stake on the urban stage.

Second, this domination operates a social selection and hierarchization of urban populations by valorizing the presence of some professional groups to the detriment of others. A categorization of uses and users therefore takes place. This differentiated management of urban space and of the social groups therein is justified through reference to the notion of the trickle-down effect, according to which this concentration of public investment in strategic areas produces benefits for all concerned. According to the same principle, an 'ecologically correct' habitus and 'ecologically correct' behaviours are similarly destined to spread from one place to the next, from gentrified neighbourhoods to working-class neighbourhoods, where social housing stocks are the dominant feature. In fact, it is a case of progressively sanitizing and disciplining certain types of use and user of the city in the urban space, of aligning behaviour with standards that are generally recognized as legitimate and which guarantee well-being, the integrity of the living environment, and supposedly better access to the city's resources. These standards are generally those most widely shared by members of the creative class and those belonging to educated and privileged milieus (Rousseau 2008). It is a known fact that it is the representatives of the upper middle classes who express social demands in terms of quality of life. In other words, these 'social circles' are tremendous assets for attractiveness policies. The effect of this process relates to one of the fundamental drivers of cultural domination as described and analysed by Pierre Bourdieu. It should be noted that such strategic objectives set down by cities and other local authorities—either individually or in partnership—are generally displayed under a banner and rhetoric which reflect consensual intentions based on the attractiveness, sustainability, and social cohesion triptych. At the risk of repeating myself, let me say that 'low-carbon and inclusive' urban development has, in recent years, become a must on the policy agendas led by cities in France and in Europe. It has to be said that it is also one of the ideological and practical pillars of the European Union's urban development programmes.

The quest for a good business climate and for neighbourhoods to seek a low-carbon trajectory is thus accompanied by the implementation of mechanisms for the social control of the space and of the populations therein. While this objective of becoming cleaner can be direct and explicit, it can also be more insidious and diffuse through the promotion of 'good individual behaviour' inspired by guidelines, 'good practice' catalogues, and benchmarking approaches which are produced and propagated via the ULLs' current activities. It is important to remember that the latter constitute the building blocks of a soft governance system that aims to include the uses and requests of residents in an urban offering which combines the objectives of economic growth, social justice, and ecological transition. Additionally, they aim to make it easier to take into account the needs and requests of inhabitants in order to co-construct programmes and improve local democracy. Faced with the scale and universal nature of this project, one can only bow one's head, as dissenting voices have difficulty in making themselves heard.

3.4 Gramsci and the ULLs

We should stress that ULLs and other instruments of empowerment and participatory activism also have a moral dimension and a cultural impact on individual awareness. They represent a brick or a constitutive part of a hegemonic regime in the sense of Antonio Gramsci: they legitimize a model of urbanity made up of good users and good uses and establish the domination of certain social groups and urban coalitions according to renewed, diffuse, and subtle mechanisms. Directly or indirectly, this vision of hegemony follows Pierre Bourdieu's approach to cultural domination as a key element of social reproduction (Ciavolella 2020).

In the current context, ULLs are vectors for the transmission of new responsibilities and obligations towards Mother Nature and Planet Earth. When we look at the aims of many of the ULLs that have been created in recent years with the help of national or European subsidies and which are managed by actors, academic or not, we find that they are imbued with a 'superego' in which worldviews and practices in favour of the environment and the fight against climate change generally dominate. They aim to disseminate an ecological ethos among groups of users or residents. At the same time, they also blithely dream of an inclusive society and the social integration of populations suffering from marginalization. This initially concerns gentrified neighbourhoods whose residents find themselves being awarded prizes for excellence in urban sustainability, but this sobriety and the moral pressure which accompanies it also tend to spread to working-class neighbourhoods. Urban regeneration policies do not escape this rising tide of standardization on the road to the ecological conquest of the zones prioritized by city policy. In these neighbourhoods, where they exist, social ULLs generally operate on two levels, mixing the goals of social justice and environmental justice. They adapt social compensation actions to the ecological label, so to speak! Today, one cannot imagine an inclusive city that does not simultaneously stand to attention as a sign of respect to 'its' Majesty the environment. In this regard, experimental approaches are suitable test beds for trying out new incentives designed to spread 'low-carbon' or ecologically virtuous behaviour. From this perspective, ULLs play a normative role that encourages the empowerment of actors so that they integrate a habitus of, if not minimization, at least moderation of their environmental impact or carbon footprint. This process reflects a slogan addressed to city dwellers which often appears on the guideline leaflets, booklets of recommendations, or best conduct 'bibles' created and disseminated by the social and ecological entrepreneurs of cities in transition. Such catalogues challenge the city dweller as a subject and, on occasion, dictate rules of conduct in a *vademecum* intended to shape the reference image of 'a responsible citizen of the sustainable and inclusive city'.

It is worth noting that empowerment is an activator of individual responsibility for taking one's fate into one's own hands and becoming an entrepreneur of one's own life, and is a psycho-social process of integrating values designed to 'green' social behaviour. The scholarly doxa of the Anthropocene and the rising influence of 'fashionable' ideas are a sure sign: it is necessary to internalize the rules of a discipline, if not to cherish Nature then at least to spare and respect—to revere?—its Majesty the environment! Is this not a supreme path, staked out by what we might call 'bio-ULLs', leading to 'ecological empowerment' and the acquisition of an environmental conscience? This is somewhat hazardous, but it might be effective if prototypes, good practices, or other suitable references are 'shaped' (ecological design) with a healthy pedagogical capacity to spread the ecological word. We have to accept that today, citizens or consumers must know how to—and learn to—cultivate their shared gardens and monitor their carbon footprints. In this atmosphere and this historical moment, the roads of bio-power analysed and denounced by Michel Foucault are not far away… Similarly, they echo a slippery slope that heralds a possible ecological dirigisme, a new moral order, and a road to servitude restricting the field of individual liberties (Hayek 2013). This arbitrariness generated by an all-powerful force, radicalized in its expressions and modes of operation, has found a striking and tragic illustration in the health policies that were pursued, particularly in France, in 2020 and which continue at the time of writing.

Why make things more complicated than they need to be? A carbon tax, a price signal, would be infinitely more transparent and effective for encouraging respect for the environment. Here again, the market can provide free-choice solutions that do not require mechanisms that moralize social life. In other words, can we count on the neutrality of prices and the market and stay away from the programmes of social and bureaucratic entrepreneurs who want to reform society by pretending to improve it? No, because in the current circumstances, the ideological power of fashionable ideas excels in creating tools to serve ideals (the sustainable and inclusive city) that are supposed to guarantee the achievement of the common good. This represents a rising force that flattens and levels awareness by contributing to new forms of 'bio-empowerment'. The urban ecological order and that of the inclusive city is thus deployed by consent, through consensual alignment with the dominant and diffuse representations that impose behavioural norms by transforming individuals into 'good city dwellers'.

In this spirit, there are different levers for achieving this, and ULLs are able to apply a maieutic and an art of persuasion that financial incentives do not possess. Through these instruments, social entrepreneurs of all kinds are at work to help moralize and normalize social behaviour. In so doing, they—like it or not—contribute to a form of depoliticization of the issues on the urban scene and agenda. Thus, in the experimental city (a testing ground for 'soft governance'), ULLs are instruments and levers to support the fight against climate change. They stand out as new tools of persuasion—of domination?—and social control that aim to encourage the ecological empowerment of inhabitants. At the same time, they extol the virtues of the inclusive city and of the fight against exclusion and other forms of marginalizing fragile populations under the universal banner of social justice. De facto, ULLs constitute full-scale tests, demonstrators, and experimental prototypes to disseminate 'good practices with low environmental impact' and 'ecologically correct' beliefs in the uses and behaviour of inhabitants. Yet nowadays, social solidarity has also—and perhaps above all—become ecological and environmental, and vice versa. It is mixed up with the various vulgates and commonplaces of the resilient and sustainable city. This is why, when they exist, the ULLs located in the neighbourhoods of large social housing complexes participate in this cultural hegemony, this levelling of awareness, and this standardization of behaviour and representations in the name of the neoliberal credo that the 'inclusive and sustainable city' represents.

References

Boltanski L, Chiapello E (1999) Le nouvel esprit du capitalisme. Gallimard

Bulkeley H, Castan Broto V, Maasen A (2011) Governing urban low carbon transition. In: Bulkeley H, Castan Broto V, Hodson M, Marvin S (Eds), Cities and low carbon transition. Routledge, pp 29–41

Bulkeley H, Marvin S, Voytenko Y (Eds) (2019) Urban living laboratories: conducting the experimental city? Europ Urban Regional Stud 26(4):317–335

Canapero L, Benavero F (2016) Neighbourhood regeneration at the grassroots participation: incubator co-creative process and system. Int J Archit Res 10(2):204–218

Ciavolella R (2020) Les deux Gramsci de l'anthropologie politique. Condition humaine/Conditions politiques, 1. https://revues.mshparisnord.fr/chcp/index.php?id=109

Douay N, Prévot M (2016) Circulation d'un modèle urbain 'alternatif'? Le cas de l'urbanisme tactique et de sa réception à Paris. Echo Géo 36. https://doi.org/10.4000/echogeo.14617

Epstein R (2006) Gouverner à distance: Quand l'état se retire des territoires. Editions Esprit

Hayek FA (2013) La route de la servitude. PUF, Quadrige

Marvin S, Bulkeley H, Lindsay M, McCormick K, Voytenko Palgan Y (eds) (2011) Urban Living Labs. Routledge, Experimenting with city futures

Reigner H (2013) Sous les pavés de la qualité urbaine. Gouvernement des territoires, gouvernement des conduites et formes renouvelées de la domination dans la ville néohygiéniste [Habilitation, Institut d'Etudes Politiques de Paris]. https://tel.archives-ouvertes.fr/tel-00842287

Reigner H (2015) Une gouvernementalisation néohygiéniste et néolibérale des conduites. La ville durable controversée. Les dynamiques urbaines dans le mouvement critique. Pragmatismes

Rousseau M (2008) Bringing politics back in: La gentrification comme politique de développement urbain? Espaces et sociétés 1:132–133

Roux E, Marron Q (2017) Les Living Labs, de nouveaux dispositifs d'action publique pour penser les métropoles et les territoires. Canad J Regional Sci Canad Regional Sci Assoc 40(1):33–41

Scholl C, Kemp R (2016) City Labs as vehicles for innovation in urban planning processes. Urban Planning 1(4):89–102

Scholl C, Kemp R, de Kraker J (2016) City Labs as instruments to shape common ground on urban sustainability [Symposium paper, ICIS Maastricht University]

Severini F, Salet W, Markins M (2015) Amsterdam: regulation dilemma in IJburg and Overamstel. Aprilab, Regulation dilemma report. https://pure.uva.nl/ws/files/2520878/163696_aprilab_regulation_dilemma_report_final_deliverable_3a_1_.pdf

Smas L, Schmitt P, Perjo L, Tunström L (2016) Positioning ULLs—a new form of smart governance? In: Schrenk M, Popovich VV, Zeile P, Elisei P, Beyer C (Eds), REAL CORP 2016—SMART ME UP! How to become and how to stay a Smart City, and does this improve quality of life? Proceedings of 21st international conference on urban planning, regional development and information society, pp 919–923. https://repository.corp.at/179/1/CORP2016_112.pdf

Swyngedouw E (2011) Les contradictions de la ville post-politique. a la recherche d'une politique démocratique de production environnementale. In: Béal V, Gauthier M, Pinson G (eds), Le développement durable changera-t-il la ville? Le regard des sciences sociales. Publications de l'Université de Saint-Étienne, pp 129–159

Von Wirth T, Fuenfschilling L, Frantzeskaki N, Coenen L (2019) Impacts of urban living labs on sustainability transitions: mechanisms and strategies for systemic change through experimentation. Eur Plan Stud 27(2):229–257

Chapter 4
Urban Living Labs: Insights for Institutionally Promoted Urban Policies

Alice Selene Boni

Abstract This chapter presents the results of a study that analyzed the conditions in which it is possible to scale-up to the Urban Living Lab (ULL) approach, which was developed in large-scale social housing neighbourhoods, characterized by phenomena of social and spatial marginalization. Specifically, the aim of the study is to provide indications and tools, but also indicate challenges and critical issues to those public institutions interested in promoting and adopting, in such contexts, the ULL approach for programmes of social and/or urban regeneration. The study is based on a comparative analysis of local experimentations produced both by launching or consolidating the ULL, promoted by universities within the SoHoLab project in three European cities and by a public programme developed in the Lombardy region from 2014, which has features in common with the ULL. The study analyzed the documentation produced and thirty interviews with different stakeholders (public institutions, NGOs, social housing companies). Considering some specific topics and the important lessons learned from the local projects, the chapter deals with the theme of the transferability of labs, from a local level to a regional or national level, by an institutional lever.

Keywords Local experimentation · Scaling up · Policy design

A. S. Boni (✉)
Milan, Italy
e-mail: aliceseleneboni@gmail.com

N. Aernouts et al. (eds.), *Urban Living Lab for Local Regeneration*,
The Urban Book Series, https://doi.org/10.1007/978-3-031-19748-2_4

4.1 Introduction

The aim of this chapter is to discuss the conditions for the scaling up of the SoHoLab project.[1] The main aim of this European project was to learn how and in what conditions it could be possible to apply and develop a Living Lab approach in large-scale social housing neighbourhoods characterized by phenomena of social and spatial marginalization. The issue of scaling up means, therefore, to think of the possibility to transfer the same programme/approach or experimental policy realized in a specific territory to other contexts, especially after some elements of success have been observed. This issue is considered central (Hartmann and Linn 2007; European Network of Living Labs 2020; United Nations Development Programme [UNDP], 2013), particularly by those public institutions who promote programmes similar to SoHoLab at different levels (from the European to local level) and that are characterized by an experimental and a place-based approach. The presence of these two features on the one hand represents an opportunity for the institutions to learn and collect important information (in particular knowledge and teachings) about how to design a policy; on the other hand, it presents them with a double challenge that sounds like a dichotomy: to arrive at the modelling of the intervention programme (going beyond a preliminary experimental programme) while ensuring that this programme is, as much as possible, appropriate for the context in which it is proposed.

Regarding the first challenge (modelling), public institutions usually have the task of advancing from an experimental dimension realized at a local level to a policy design that can be extended, in a generalized way, to other contexts. This leads to the issue of the standardization of some experimentation elements which are reaped from specific contexts characterized by some peculiarities and different complexity levels. Even if the programme might generally concern large-scale social housing neighbourhoods characterized by phenomena of social and urban marginalization, it is possible to consider different factors. Thinking specifically about territorial differences, this could include the presence of different needs, resources, competencies, actors and social/political/cultural stories that may uniquely characterize these places. The standardization of public services and programmes is an ever more present problem for institutions that must guarantee equal access to services and at the same quality. Therefore, standardization means avoiding territorial inequalities of access but also, as is common knowledge, defining a problem management strategy. This is particularly important in large-scale contexts (European, national, and regional) to simplify government activity and produce cost savings and restraints. The standardization of services, in fact, enables better control of public expenditure but also expected results, by the definition of some general indicators. Speaking about standardization and modelling means also, in general, considering the simplification process of social complexity that is required to manage social and spatial problems from an administrative point of view (Tosi 1994).

At the same time, for public policies to be effective they must be adequate for the specific territorial contexts in which they operate. This second challenge is central.

[1] For more information about the project see the Introduction of this book.

In the premises of the SoHoLab project, one of the main aims was precisely to understand how it would be possible to adapt top-down regeneration programmes to neighbourhood needs using the Urban Living Lab (ULL) approach. The project, in fact, started from the evidence that part of the failure of many regeneration programmes was due to the top-down approach used by institutions that, in a certain way, compelled the territory to adapt itself to the characteristics of the programme, and not the contrary (Aernouts et al. 2020).

Therefore, in short, the problem seems to lie in how to satisfy the institutions' needs to promote a generalized policy, while at the same time protecting the necessity to respond to specific local needs. This second challenge, which appears to clash with the first (modelling), is more complex to treat. It deals with how to design services and programmes tailored to territories, with social complexity management and, specifically, the possibility to integrate and extend a place-based approach not only in experimental programmes but in generalized programmes as well. In this chapter, an effort will be made to understand how it is possible to combine the tendency for standardized programmes and the definition of an intervention model with the need to design a project and programme tailored to the neighbourhood characteristics. For this purpose, in the next paragraphs, the results of a comparative study[2] are presented. The experiences of applying the ULL approach in public housing contexts, carried out as part of the SoHoLab project, and a similar programme, called Neighbourhoods Social Lab (NeSoLab, Laboratori sociali di quartiere), promoted by the regional government in Lombardy (Italy), were compared.

4.2 The Interpretative Dimensions of Analysis Examined

The study was carried out through the use of qualitative methodologies (documentation analysis and interviews). It was realized in the last nine months of the project and organized into two phases, which were linked to two different comparison grids. In the first phase, an analysis of the documentation produced during three years of the project was completed. The purpose was to collect and compare the general characteristics of the programmes in terms of the main characteristics of the territories selected for the intervention, typology of involved actors, problems confronted, objectives, main activities and project typologies realized, the different origins of the resources, and project development period. The outcome of this part is presented in the next section.

In the second phase, eighteen semi-structured interviews were conducted that allowed us to collect the perceptions and points of view of the main stakeholders involved in the two programmes (in total twenty-four people including public institutions, NGOs, social housing companies). The grid used was designed to compare and

[2] This contribution is based on the results of the research carried out by Marcella Bonanomi and Alice Boni between April 2020 and December 2020, as part of their research activity in PoliS-Lombardia, local partner of the European project SoHoLab (Bonanomi and Boni 2021; Boni 2021).

deepen some specific interpretative dimensions with regard to the main characteristics and purposes: the key aspects of the ULL approach but also most urban and social regeneration programmes. Specifically, the dimensions considered were: participation, learning processes, siting, integration among sectoral policies (social, educational, town planning, etc.), integration among different levels of actors (different institutions, local realities, citizens…), the time factor, changes, and legacies. The results of this part are presented in Sect. 6.4.

Taking into consideration the double challenge presented in the introduction (to reproduce from a top-down level a ULL approach in any context without losing the complexity gained from a place-based approach), each dimension was investigated with two aims. Firstly, the aim is to understand the implications of its implementation, from the theory and aims expressed in the initial projects to practice. Secondly, with the aim to flesh out attention points and potential recommendations for those institutions that decide to adopt a programme with a similar approach. Considering the first dimension, *participation*, the different senses, meanings, and expectations assigned by institutions to the residents' activation and involvement in the programmes were investigated, also in terms of the role conferred to them in the process. For example, if participation is intended as being an apparatus *(dispositif)*, institutions' role is pedagogic in order to activate an attitude of responsibility for taking care of the public space in residents; their role is fact-finding in order to involve residents to better know their needs, problems, and competences and design more appropriate responses. This dimension was also investigated with reference to the characteristics of institutions' and social organizations' participation in the two programmes, in particular with the aim to study in deep of their role and the importance of their involvement in the process. The second dimension regards the *learning process*. Given that the two programmes were characterized by a local and place-based approach, efforts were made to investigate if the public institutions pay heed to the local and scientific knowledge acquired produced thanks to these programmes, and if and how they transmit and use this knowledge at the highest level to define or redefine a policy. The third dimension concerns *siting* and the first aim was to investigate it understanding the different meanings given to the 'institution' presence in the neighbourhood. In particular, did the presence of the institution in the neighbourhood mean greater social control of the space and its uses, or the possibility to involve residents in management activities? Did the 'institution' assume a different gaze (immersing himself in the reality of the neighbourhood thank to his physical presence), and consequently, receive different knowledge to manage but also to design the policies and build local relationships? This dimension seems strongly linked to the meaning assigned by the institution to the participation concept. Secondly, the aim was to investigate the different strategies used to stay in the field, to understand whether there are methods or physical spaces that are more effective for a specific purpose and the different possibilities to be present in the neighbourhood. Another dimension that is peculiar in the programmes examined concerns *the integration between policies*. Both SoHoLab and NeSoLab start from the assumption that social and urban problems in large social housing neighbourhoods are strongly interrelated and that a regeneration programme, to be effective, has to conceive any policies in an integrated way. If on the one hand, the regeneration programme experiences of the last twenty years resulted in these lessons, then, on the other hand, putting the integration into practice is a difficult challenge; in fact, it is observed how generally the problems are faced independently of each other (Aernouts et al. 2020). The aim of this part was to understand what are the main difficulties, challenges, and problems related to the realization of an integrated regeneration programme. Strongly linked with the previous dimension is the *actors' collaboration*. Here, the aim was to try to investigate the integration among different levels of actors involved, with different perspectives, in the neighbourhood's life: public institutions (regional, municipal, etc.), local organizations, and residents, in order to understand the presence of specific apparatus used to foster integration among different actors and to try to identify the main critical issues and challenges for integration. With the investigation of the *time factor*, an attempt was made to understand whether the actors (who promoted the two experiences examined) have considered this dimension with respect to the definition of their aims and the expected impacts from the implementation of the programs. In particular, the goal was to comprehend if the interventions, activities and strategies had been designed in the programmes to be limited in time, inasmuch they had to be strictly useful to achieve some specific objectives. Or, quite the opposite, if they had been designed to continue in time, inasmuch they were conceived to transform, after the experimental phase, in a model of management or in a consolidated and structured activity in the neighbourhood.

Finally, in both projects considered, specific attention was paid to the need to produce *changes* in large-scale social housing estates through an urban regeneration process based on the ULL approach. This is an objective that institutions from different levels (European, regional, and local) are responsible for, in particular in the premises of the projects, where the expectation is to produce visible and evident changes. This particular attention is also caused by the fact that often, regeneration programmes of public housing neighbourhoods work in contexts considered static and passive, where public intervention does not seem to have the power to produce changes. By investigating this dimension, the aim was to understand the main difficulties and challenges that actors have to deal with to produce some effective changes; the realistic changes that institutions can expect in terms of visions, political frameworks, political practices, and spatial and social aspects; and finally, strongly linked with changes, the possible main *legacies* of these kinds of programmes.

4.3 The Experiences Examined

The two programmes examined in the comparative analysis represent an experiment that was developed between 2017 and 2020. The general area that unites them is the urban and social regeneration of public housing neighbourhoods characterized by phenomena of urban and social marginality. Among the common objectives, we find a focus on practices, processes, and tools to promote social cohesion and spatial inclusion, in particular: listening to residents and encouraging their greater involvement in the management of the neighbourhood, rebuilding a relationship of trust between residents and institutions, and reconnecting the neighbourhood, its residents, and the city. Further similarities can also be identified in the 'model' of intervention, with reference, for example, to the importance given to certain elements such as the involvement of residents, the construction of collaborative practices between different institutional actors involved in neighbourhood life, the construction of a network of local actors, and fieldwork.

However, the two programmes arise within different institutional frameworks. The NeSoLab is an experimental programme promoted within the housing policies of the Lombardy Region that has the definition of a model of management of public housing as its final objective. This model must be able to, according to the idea of the legislator, integrate *property* and *facility management* with social management (currently absent and tested through NeSoLab). The latest edition of the regional programme was funded by European resources from the European Social Fund (ESF) 2014–2020. The ULL experience developed within the SoHoLab project was born, instead, within the framework of European programming from an idea by three universities with the aim of promoting an integrated approach to the regeneration of public and collective spaces of public housing neighbourhoods. This would be done through a retrospective analysis of similar past projects carried out in the Paris metropolitan area, a ULL experience underway in Milan, and a new project of ULL in Brussels.

Although the two experiences were born within different institutional contexts, the numerous common elements were the basis of the idea to develop an in-depth study aimed at identifying teachings and indications for the reproduction and transferability, through an institutional path, of analogous programmes in areas with similar characteristics and problems.

4.4 Points of Attention for the Adoption by Public Institutions of an Urban Living Lab Approach

This section presents the results of the in-depth analysis that was conducted of each of the interpretative dimensions described in section *The interpretative dimensions of analysis examined.*

4.4.1 Participation and Role of Actors

Residents. Both the SoHoLab and the NeSoLab programmes provide some important food for thought with respect to urban and social regeneration processes that aim to involve residents. First, the importance of a preliminary activity aimed at listening to and knowing the needs of the population in terms of housing and social and living conditions was underlined. Second, it emerged that the participation of residents, with this term indicating the tenants who live in a social housing complex, is almost never attainable. It is therefore worth starting from the position of focusing on certain population groups, which can be useful as well as more realistic, especially in contexts with a large number of residents. In order to be able to involve residents, it seems important to identify the problems which they consider to be their priority together with them. Doing so using predefined themes and programmes risks making engagement attempts unsuccessful. Under these first conditions, participatory processes must be conceived as experiences within which residents can contribute to real change; in this sense, it is important that their issues do not remain disregarded. If, therefore, the demand for greater involvement on the part of residents is not associated with an equal effort on the part of the institutions, the objective of greater participation is at risk of disappearing. The experience of the NeSoLab has shown in particular how programmes that aim to involve residents can be interpreted differently at the local level in relation to the sensitivity of the actors present, who may associate participation with a multiplicity of meanings. It is considered useful when writing calls and programmes or sharing them with local stakeholders to explain the purposes of so-called participation. One of the possible distortions that can be triggered is the result indicator represented by the number of residents who have been involved in the initiatives as a measure of their effectiveness. The risk is that mechanisms will be found in which counting the presence is the ultimate aim instead of reaping the real transformative potential of the participatory processes.

Institutions. The institutions involved in the two programmes, from government agencies to social housing companies, have played a crucial role in determining the outcomes of the processes, both positively and negatively. This was due to a number of factors. First, the sensitivity of the people directly involved in the projects on behalf of the administrators and officials. Second, the involvement in the project of the entire organization structure (from the top to the most operational levels) to which these people belong. Third, the possibility of having resources (human, technical as well as economic) and power of action to be made available in the project in order to respond to issues (new and unexpected) that emerged from the dialogue with residents.

Starting from some critical issues that emerged but also from the most successful cases, for all stakeholders we tried to understand what skills and conditions the institutions that are managing, promoting, or experimenting with urban and social regeneration programmes within public housing neighbourhoods should have and to be able to put into practice. These include, in addition to technical and administrative skills:

- listening skills of local stakeholders, workers operating in the territory, or tenants, taking into account any requests and trying to respond to them;
- the ability to be in the field, through the opening of a location;
- the capacity to work in synergy with local networks and other institutions;
- the ability to build settings in which to experiment with innovative solutions, which are in addition to the ordinary intervention procedures;

- to create an internal culture within the organization that is favourable to participatory processes;
- the ability and possibility to work in a coordinated way between different sectors to meet the complex needs of citizens;
- the ability to inform and communicate with citizens in a transparent, coherent, and coordinated way with other institutions;
- know of the territory in which it is operating;
- the ability to employ a broad perspective on the processes and, at the same time, segment the vision into practical tasks and concrete actions.

Universities and third-sector organizations. In this section, it was decided to look at the role of the universities and third-sector organizations together because their work has, in some respects, been similar. It was therefore considered useful, with regard to the possibility of making the two experiences reproducible and transferable, to better understand the peculiarities of their contribution.

Both the university in the SoHoLab project and the organizations of the third-sector in many of the NeSoLab projects were central actors in the various phases that distinguished the programmes: from the design programmes to the ones dedicated to the knowledge of the context and the study of residents, to those of management and coordination, to those of implementation of the actions, to the monitoring and evaluation of experiences and dissemination of results. They played an important role in mediating between the actors involved and in building or strengthening the network of local actors.

However, this was done within different roles, institutional frames, and assignments. As pointed out by the researchers themselves, in the projects the university took on the role of intermediary; a broker with an independent perspective (Maranghi et al. 2020, p. 64). This was possible because the three universities were not directly dependent on any of the actors involved in the project. However, the possibility of playing that role within the contexts existed in part thanks to the authority that distinguishes the institution (facilitating some processes and making it possible to open some communication channels with other institutions) and, in part, thanks to the effective recognition of this role by the other institutional actors that depended both on the availability of them and on the skills and competences of the researchers.

Third-sector organizations, on the other hand, often assumed the role of a social manager within a relationship that saw them being closely dependent on one of the two most important public institutional actors in NeSoLab: the municipality or the social housing company. While in some projects this condition effectively legitimized their work, in other cases this was a weakness, especially where the relationship between institutions and residents had particularly deteriorated (for example in terms of lack of trust in institutions). The effectiveness of the action of these organizations therefore depended both on the skills and history matured in the neighbourhood and on the role and authority attributed to them by the various institutional actors. Their ability to influence contexts was often directly proportional to the attention, care, and investment that institutions put into the process. In cases where the institutions totally delegated to the animators the relationship among the residents (making them, in actual fact, the institutions' terminal), leaving them without resources and power to act, their role lost its importance and effectiveness.

The position of the third party seems to be important because it allows actors (from residents to institutions) to be approached outside a relationship of dependence and to operate in the process, bringing their own perspective that is not necessarily that of the institutions nor that of residents. This can help foster the construction of more constructive relations where there are tensions or conflicts between different institutions, or between residents and institutions. Second, it can ensure that processes and their analysis are unhindered by the influence of contractors. However, the possibility of taking on this 'third' role depends on certain conditions that are difficult to reproduce: the possibility of drawing on funding channels that do not depend on the institutions that are part of the network, and the recognition (as has been pointed out earlier) of the network of actors within which action is taken. If this is not the case, it is important that the organizations entrusted with carrying out the tasks of design, management, coordination, and implementation (and social management) do so within collaborative processes where all institutions feel strongly involved.

In general, in the opinion of some public administrators, the presence of independent intermediaries or brokers should be limited to triggering and consolidating a more direct relationship between residents and institutions. This is a perspective that can be shared by other stakeholders who, however, emphasize that in contexts of public housing characterized by the presence of situations of strong social and urban marginalization the capacity for self-organization and vocal participation of residents takes a very long time to be realized.

One aspect that concerns universities is their ability to provide a reflective analysis that represents an added value to the public housing neighbourhood but is considered difficult to reproduce within the programme.

Apart from the question of independence, from both experiences, it has been possible to identify the profile of those who can play the role played partly by the university and partly by third-sector organizations in similar experiences:

- local actors, who know the place in which they intervene;
- subjects with specific skills (such as design skills);
- subjects able to understand people with different backgrounds with a fine-grained knowledge of human needs and cultural dynamics;
- subjects with a strong reflexive capacity;
- subjects able to be mediators and activate the local community.

4.4.2 Learning Process

The experience of SoHoLab and that of NeSoLab shows that the transmission of locally produced knowledge at the highest levels (in this case the institutions) does not take place automatically and linearly.

The conditions that favour this step seem to concern, first, the ability of the organization that facilitates the process to translate and effectively return results to the highest institutional levels, and, again, strong involvement on the part of institutions. This last aspect implies that institutions have:

- availability in terms of time, in order to follow up on the lessons learned (following established practices and procedures requires less resources and energy in the short term);
- an interest in staying and actively participating in the process. This interest should affect the whole organization from the outset, so that the person who participates in the process is entitled to transfer knowledge learned within the organization and in particular to the decision-making levels so that they can choose to use it.

The risk that this rich wealth of knowledge will remain unused by institutions is very high in all the experiences analysed.

4.4.3 Siting

Presence within the neighbourhood represents a practice considered fundamental for both programmes because it makes it possible to:

- build meaningful, trusting, and trustworthy relationships between actors, both at the institutional level and with local actors;
- access different forms of knowledge;
- understand the neighbourhood from the inside, instead of from the outside;
- involve residents more in action research and, at the same time, have greater opportunities to be involved by local actors.

The implementation of this presence, as suggested in particular by the experiences that have developed in the SoHoLab project, cannot be predefined but requires the most appropriate mode to be chosen in view of some aspects that concern both the conditions laid down by the project or programme in which it is inserted (the objectives and the duration of the stay that depends on the resources available) and the characteristics of the context (the dynamics that invest it and the role that it intends to play with respect to the network of actors who live there). These assessments seem necessary in order to create the conditions for the construction of a positive path of legitimacy and recognition by the institutions but above all by residents, who find themselves welcoming ‘new tenants’ into their territory.

At the same time, it has been pointed out that, in the event that the opening of a space is chosen, it is important to confront the time and resources factor. The reactivation of a place, although viewed with suspicion at an early stage, often represents a lighthouse that illuminates a dark and abandoned space, a symbol of a phase of abandonment and disinterest on the part of the institutions. The care taken of the space and the investment of resources in reusing it, is the first sign of the desire for change on the part of institutions. At the same time, its closure at the end of the project, from a symbolic point of view, can suddenly break a laborious process of building a relationship of trust with residents and can be a further reason for their departure from institutions, making it more difficult to involve them with another project.

4.4.4 Policy Integration

In general, both experiences have stressed the importance of actions that are as integrated as possible, due to the need to provide and build interventions capable of responding to complex problems. For integration to take place positively, a number of conditions that concern institutional ability seem necessary. These include the ability of institutions to:

- analyse interdisciplinary as well as intersectoral problems;
- know how to reorganize themselves internally to enable the various sectors to communicate as much as possible;
- work together among actors operating in different sectors;
- gather the knowledge that is necessary by activating professionals and actors with skills that can help fill information gaps;
- streamline administrative procedures, so that decisions taken through interaction between the various actors are feasible and effective.

Another important condition regards the availability of resources to invest in the project. As part of the programmes, references to the cross-sectoral nature of the interventions also paid close attention to the weak investment in recent years in the quality of physical space, which contributed to the generating an important reduction in the physical quality of public housing complexes, both with regard to housing and with reference to the common parts. An important point of attention that has been highlighted by the actors interviewed is the integration between the social and spatial dimension of the interventions. In particular, it was stressed that a balance needs to be restored between, on the one hand, the focus on the process, the way in which the network of local and institutional actors operates, and the involvement of residents and, on the other, interventions in the redevelopment of the physical space. Experience has shown that the effectiveness of these projects can be undermined if the only objective is social compensation and the fight against poverty.

4.4.5 Actors' Collaboration

The experiences of collaboration between different organizations and institutional levels analysed in both programmes confirm the importance of these processes and the values and advantages associated with them in terms of the effectiveness and appropriateness of the solutions identified. Unity of purpose, the sharing of objectives, and collaboration while working towards their achievement seem fundamental, and the failure to commit by some institutional actors weakens the project and renders it ineffective. The main issues associated with integration between the actors were the following:

- From the point of view of the relationship between local institutions and organizations and residents, it is important for institutions wishing to create networks of collaboration between actors to include the most appropriate tools for recognizing roles and decision-making power within collaborative processes.
- From the point of view of the relationship between the different institutional levels, it is necessary to establish increasingly collaborative working practices that overcome competitive or political dynamics.

4.4.6 Time Factor

The experiences of both SoHoLab and NeSoLab have highlighted the importance of urban and social regeneration processes being included in a long-term planning strategy based on the existence of certain and continuous resources, which exceeds the experimental aspects that often characterize them.

First of all, time is necessary in order to get to know the context, to involve the actors, to trigger processes of empowerment of residents, to learn a different way of relating, and to create those assumptions that make it possible to coproduce and comanage projects but also take care of the network. The end of the project is an element of concern with regard to the energies and resources mobilized, and the greatest fear is linked to the risk of nullifying the efforts made by triggering counterproductive processes, both with regard to the expectations created in the actors involved and to the waste of resources that would result.

The crucial question therefore concerns the possibility of finding resources (human, social, or economic) so that the continuity of the presence is guaranteed as long as it is necessary for the purposes of the project and the objectives set. However, it is strategic that short-, medium-, and long-term objectives are set to avoid fuelling management processes that no longer look to solve problems but instead aim for the survival of the apparatus created.

4.4.7 Change and Legacies

Producing a change in portions of cities characterized by phenomena of urban and social marginality that have consolidated over time is generally viewed with distrust, both by their residents and by institutions. With regard, therefore, to the theme of stagnation that often seems to characterize the narrative about these neighbourhoods, the experience of the researchers of the SoHoLab project suggests broadening the view to include not only the changes that we expect but also those that really take place in parts of the territory that are not at all stagnant, for example when we consider the social practices that take place in them. In addition, SoHoLab's experience suggests reflecting on changes considering not only the neighbourhood and the people who live there but also how it is studied, inhabited, and represented, with reference to the tools, sources, or methodologies that are used to do so (and that often do not allow some transformative processes to be read and interpreted) and to the actors of change, which could be individual residents or institutions.

Both the projects in the SoHoLab programme and those of the NeSoLab constituted experimental sites that showed how the impact of the project:

- is difficult to measure, especially when it concerns a change in perception or vision that can concern both a plurality of actors and individual people;
- may lead to the establishment of new instruments for coordination or intervention;
- can lead to the transformation of small portions of territory;
- may have a predominantly indirect impact on residents.

It has also been pointed out that these changes may depend on the sensitivity of individuals, their interest, and their willingness to question established ways of working, but they also depend on the complexity of the issues to be dealt with. The more complex and intractable the subject, the more is the resistance of the institutions that have to deal with it.

Setting change objectives that are commensurate with the resources and time available is certainly an important stimulus for policymakers and project protagonists. However, the usefulness of reviewing higher objectives of change remains firm. This helps to demonstrate the effectiveness of the programmes already implemented, set up long-term programmes and reflect on the systemic conditions (institutional, economic, political, and social) that contribute to feeding the social and urban marginality that have characterized public housing contexts for many years.

4.5 Conclusion: Promoting Urban Living Lab Experiences in Large-Scale Social Housing Neighbourhoods by an Institutional Lever

Different dimensions of the ULL approach that can be proposed in urban and social regeneration programmes have been examined in the previous paragraphs. In particular, the implications for institutions that want to adopt this kind of approach and the conditions for its possible success were taken into account. These are not solutions suitable for every situation and territory but points of attention that could be taken into consideration in the design and management of similar programmes. An effort was made to identify possible consequences of the attribution of a particular meaning to a single dimension in order to better guide the decisional process, because it seems necessary to be aware that each dimension may be filled with different meanings for the actors.

Considering the particular role that an institution may play in these programmes, in these conclusions it is important to draw attention to some essential aspects that could allow a positive and conscious experience in the assumption of the ULL approach (and 'attitude'), but it is also important to capitalize on, as much as possible, all potential advantages given by this approach. It is possible to confirm that the ULL approach allows for an adaptation of the regeneration programme to different contexts thanks to the construction of collaborative processes between different available actors and also the activation of a 'cumulative' learning and knowledge process. From the point of view of the institution, these are undoubtedly advantages in terms of both programme success possibilities and programme effectiveness. However, what the SoHoLab project and the NeSoLab experiences demonstrated is that the process of adapting policies, programmes, and services to contexts and the activation of learning and knowledge processes are possible as long as there is a willingness on the part of the promoters (the institutions) to adopt a reflexive and dialogical approach in the territories in which the intervention 'lands', and a willingness on the part of local actors (residents and organizations) to collaborate and participate.

The participation of local actors is probable and not predictive. Instead, it is essential that the institutions have the willingness to be in the regeneration process with this openness. Institution refers to everyone who works inside an institution, from the higher and political levels to lower levels. For the institution, this willingness implies a particular attention to and care of the process and, importantly, the awareness that the programme could be modified and adjusted during the process while respecting general principles, particularly in response to events and situations that may be generated by the interaction among actors and by the continued acquisition of usable knowledge and discoveries. Events and situations could be classified by institutions as unexpected and, as such, refused, but they are actually part of the nature of the process and intrinsic to every interaction among actors with different rationalities and to a place-based perspective that deals with the territories. 'Territory' refers to a set of social, economic, and environmental aspects. So the willingness to be in the process for the institution also means to pursue a particular sensitivity to manage events with curiosity (even unexpected events) instead of a defensive approach, considering them a possible output of the programme (Hirschman 1967).

It is evident how this approach contrasts with the tension of standardizing and modelling processes, services, and programmes. If, on the one hand, the objective of defining a standardized management model seems reasonable, both from the point of view of the institutions and from the point of view of citizens, experience shows that a certain amount of flexibility, openness, and listening skills on the part of the institutions promoting these services seem necessary. The risks associated with the simple and uncritical standardization of services relate to the difficulty of adapting services to the needs of residents that, sometimes, it was seen, left the latter with no option but to adapt themselves to the services. The involvement of residents and different institutional actors should make it possible to define and redefine at a local level the practices related to the management model, in a process of dialogue and continuous listening in which the ultimate objective is not so much the definition or application of a model, but a response to some shared needs.

References

Aernouts N, Maranghi E, Ryckewaert M (Eds) (2020) The regeneration of large-scale social housing estates. Spatial, territorial, institutional and planning dimensions. SohoLab. http://soholab.org/content/2-publications/1-the-regeneration-of-large-scale-social-housing-estates/soholab_report1.pdf

Bonanomi M, Boni AS (2021) Laboratori sociali e Living Lab in contesti di edilizia residenziale pubblica: un'analisi comparata. Polis Lombardia. https://www.polis.lombardia.it/wps/wcm/connect/9d36183d-5085-47ef-a3ca-ce7d49fc3152/SOC17011_SOHOLAB_Bonanomi_Boni_laboratori_sociali_def.pdf?MOD=AJPERES&CACHEID=ROOTWORKSPACE-9d36183d-5085-47ef-a3ca-ce7d49fc3152-nyCfkQf

Boni AS (2021) Laboratori sociali and Urban Living Lab in public housing contexts: teachings and proposals of contamination. Executive summary. SohoLab

European Network of Living Lab (2020) Co-creating innovation: Scaling up urban challenges from local to global. https://issuu.com/enoll/docs/enoll-olld-report_final_web_version

Hartmann A, Linn JF (2007) Scaling up: A path to effective development. International Food Policy Research Institute. https://www.brookings.edu/wp-content/uploads/2016/06/200710_scaling_up_linn.pdf

Hirschman AO (1967) Development projects observed. Brookings Institution

Maranghi E, Cognetti F, Ryckewaert M, Aernouts N, Mosseray J, Lefrançois D, Wachter S, Labied N (2020) 5 themes from urban living lab experiences in large-scale social housing estates. Soholab. http://soholab.org/content/2-publications/8-themes-from-urban-living-lab-experiences-in-large-scale-social-housing-estates/soholab_5themes.pdf

Tosi A (1994) Abitanti. Le nuove strategie dell'azione abitativa. Il Mulino

United Nations Development Programme (2013) Guidance note: Scaling up development programs. https://www1.undp.org/content/dam/undp/library/Poverty%20Reduction/Participatory%20Local%20Development/ScalingUP_guidancenote(Jan2013)_web.pdf.

Chapter 5
Adapting the Living Lab Methodology: The Prefix 'Co' as an Empowerment Tool for Urban Regeneration in Large-Scale Social-Housing Estates

Francesca Cognetti and Elena Maranghi

Abstract In recent years, Urban Living Labs (ULLs) have acquired an ever greater resonance in the field of spatial and urban regeneration. Indeed, the promotion of a collaborative approach turns out to be decisive if one wishes to include a multiplicity of social actors in these processes, an indispensable aspect today of promoting effective physical and social transformations of the urban environment. However, which specific adjustments must a ULL make in order to be configured as a truly inclusive tool within marginalized urban areas, such as public-housing neighbourhoods, where access to decision-making processes is structurally limited? Departing from a European perspective, reinterpreted through the specific Milanese context of the San Siro district, the paper reflects on the approach of ULLs in marginalized areas: material and immaterial work platforms where different languages, knowledge, values, and visions meet through an active—even conflictual—encounter which is crucial for the promotion of local regeneration processes.

Keywords Urban Living Labs · Codesign · Urban regeneration · Capabilities · Social-housing estates

---

F. Cognetti (✉) · E. Maranghi
Department of Architecture and Urban Studies DAStu, Politecnico di Milano, Milan, Italy
e-mail: francesca.cognetti@polimi.it

N. Aernouts et al. (eds.), *Urban Living Lab for Local Regeneration*,
The Urban Book Series, https://doi.org/10.1007/978-3-031-19748-2_5

5.1 Introduction

In the last years, the Living Lab approach has been increasingly applied in socially oriented urban research agendas, as the example of the Joint Programming Initiative (JPI) 'Urban Europe'[1] shows (Bylund et al. 2020; Franz 2014, 2015). Even if the Living Lab tool is widely considered capable of promoting social innovation and participation in urban environments, the academic debate on Living Labs in urban studies remains underdeveloped (ibid.). Only recently has the methodology been applied to urban and planning fields (see Rizzo et al. 2021), especially thanks to the expansion of networks such as the ones promoted through JPI's programmes. However, it still struggles to be adapted to deprived[2] or marginalized areas of our cities.[3] Yet, the approach seems promising for tackling some of the main challenges related to these contexts. As will be argued, Urban Living Labs (ULLs) share certain common elements (Malmberg et al. 2017, p. 12) that are particularly relevant in relation to the characteristics of large-scale social-housing estates, including a more effective involvement of residents in the design and implementation of urban regeneration processes. Indeed, ULLs are a set of methods that can change mindsets, processes, and material solutions (Bylund et al. 2020). Thanks to their experimental and inclusive perspective, they appear to be useful for tackling the so-called 'wicked problems' (Rittel and Webber 1973) which typically characterize these contexts and, more broadly, deprived urban areas.

Starting from these considerations and basing ourselves on the results of the SoHoLab project, we here reflect on the conditions that need to be fulfilled to adapt the Living Lab methodology to the regeneration of large-scale social-housing estates in European cities. Embracing an action-research methodology, the SoHoLab project focused on the development of field research through the improvement of an existing

---

[1] The programme launched the ERA-NET Cofund Smart Urban Futures call within which the SoHoLab project was founded. In 2017, fifteen projects were approved in the ERA-NET Smart Urban Futures (ENSUF) call, a collaboration between JPI Urban Europe and the European Commission, to address urgent and long-term urban challenges by cocreating ideas and projects. Stakeholders in over seventy cities and twenty countries were involved in the projects and published project results in a variety of forms. They reflect on 'smart urban futures' by analysing how learning, narratives, and place development are currently practiced in European cities, where current approaches are taking us, and by suggesting new ways forward through uncovering the added value of their results and integrated approaches. About half of the ENSUF projects have engaged in methods central to ULLs, with some projects even showing signs of 'Urban Living Labs 2.0'—an effort in the JPI Urban Europe community to further advance the transformation capacities of ULLs (Hawlik and Berger 2021).

[2] By using the term 'deprived areas', we refer here to the fact that these territories house a population that accumulates and faces social and economic difficulties mainly due to its institutional context (the social-housing context), such as being outside the official world of work (unemployed and retired), being a one-parent household, or having a disability.

[3] According to JPI, one of the main challenges in the development of ULLs is equity and inclusion (Bylund et al. 2020).

ULL and the implementation of a new one, both located in social-housing neighbourhoods in Milan and Brussels. In this contribution, we will mainly focus on understanding if and how user engagement and cocreation processes, which are considered central elements of the Living Lab methodology,[4] can foster the development of fragile contexts, focusing especially on the role of residents. In this regard, we will see how ULLs could function as devices in which participation is promoted through the consolidation of project platforms and the enhancement of local competences and know-how.

5.2 Fostering Capabilities as a Device for Urban Regeneration

Even though social-housing estates are often characterized by social exclusion and high levels of socio-economic fragilities, they at the same time represent 'local tanks' of competences and social resources that often tend to remain invisible. This is especially true for the most 'unheard'[5] part of the population living in these estates (youngsters; women, especially with a migratory background; people in precarious economic situations or with severe health problems; elderly people, etc.). As far as this part of the population is concerned, ULLs can potentially function as 'activators' of knowledge and competencies through the development of codesigned regeneration paths. But if, on the one hand, this is a promising possibility, on the other hand, its success depends on the extent to which regeneration policies are able to recognize and support local capacities and on the conditions under which ULLs are capable of being effective in this regard.

We propose here to consider competences as a way to express oneself and access society, both in terms of socio-economic inclusion and of full participation in citizenship. So, competencies could be described as individual qualities often linked not only with personal characteristics but also primarily with an environment that allows

[4] According to the *Living Lab Methodology Handbook*, even if there is no single Living Lab methodology, a few characteristics can be identified as the core of the methodology: multimethod approaches, user engagement, multistakeholder participation, a real-life setting, and cocreation. The characteristics are defined as follows: 'User engagement: this is rooted already in the origins of Living Labs, the key to success in any activity is to involve the users already at the beginning of the process. Multi-stakeholder participation: even if the focus is on users, involving all relevant stakeholders is of crucial importance. These include all the quadruple helix actors: representatives of public and private sector, academia and people. Real-life setting: a very specific characteristic of Living Labs is that the activities take place in real-life settings to gain a thorough overview of the context. Co-creation: typically, especially in technology projects, activities are designed as top-down experiments, benefiting from users being involved as factors rather than actors. There is an increasing recognition that this needs to change so that users become equal contributors and cocreators rather than subjects of studies. The Living Lab approach strives for mutually valued outcomes that are results of all stakeholders being actively engaged in the process from the very beginning' (Malmberg et al. 2017, p. 12).

[5] In the sense of struggling to be heard and considered in policymaking.

and encourages people to employ their skills and to act. Several pedagogical theories have demonstrated that often, public policies have inhibited people from becoming bearers of 'practical' knowledge linked with an active role. The recurring risk is to establish a 'disabling' relationship between institutions and residents, particularly in marginal contexts (Freire 1970). This means that institutions often place people in a position of 'passive receptacle': they are frequently referred to as the 'target' of an intervention (a term that originated in the field of marketing but is often used in public policies as well), meaning that they share common characteristics, and are associated with a predefined set of goals to be achieved. The target is the person to whom the intervention is addressed; the person who is the 'object' of the policy. This phenomenon is more evident in social-housing neighbourhoods, where welfare systems are often linked to mechanisms of institutional dependence.[6]

On the other hand, this seems to be exactly the point: competencies are not only a personal attribute but also a process 'under construction', a sort of 'practical reason' (Nussbaum 2011) as a combined capacity resulting from a continuous reassembly of the individual's capacity and external conditions. So, we must consider whether a whole territory, as a complex system of norms, powers, and actors, can contribute to recognizing, allowing to emerge, and valorising unexpressed and diffused skills.

In this perspective, there are two nuances of the term 'competencies' we should consider. The first is contained in the term 'capability', which links the concepts of 'ability' and 'action', suggesting a vision of know-how as an ability directly related to the type of action to which it is applied. Hence, the link between ability and action emphasizes the possibility that someone can be capable in a particular field and absolutely not capable in another; it depends on the taken type of action, as introduced by Sen and Nussbaum (Nussbaum 2002; Sen 1985, 1990; Sen and Nussbaum 1993).

The consequence of this statement, like the previous one, is still 'relational'. One holds a competence if it is useful for improving or developing a situation. Thus, competence depends on the environment in which it acts and on the person who holds it. In this perspective, not only does the 'effective' action itself demonstrate the existence of a consequence, but the degree of public or institutional formalization and recognition of the competence is also one of the elements that helps it to emerge and be properly developed. Moreover, there are often 'implicit' competencies (e.g. interpersonal mediation, language skills, leadership, etc.) that are not fully recognized as 'useful' competencies even by those who possess them. Competence therefore emerges and develops also in a mutual act of recognition. Often, the skills possessed by fragile and insecure populations are not recognized. It is therefore important to work on their emergence, 'formalization', and reinforcement.

[6] Institutional dependence is for example related to social services and access to housing and linked to a strong bureaucratization of interventions. These aspects are accentuated by the concentration of socio-economic fragility and the intersectionality of the causes of these fragilities. In these contexts, the combination, settled in space, of norms and assistance systems, personal profiles, life trajectories, concentration, and isolation makes it difficult for people to take an active role and therefore to develop and implement their skills.

Second, competence is also related to a process reinforcing citizenship, meaning the ability to act and actively participate in urban creation and urban life (Holston 1998; Isin 2009), which is usually weakened in large-scale social-housing estates. These areas are characterized by 'weaker forms of citizenship': 'traditional citizenship' does not allow many people to be represented and to have a voice in collective matters. Moreover, these places are characterized by a high presence of 'unnamed figures' (Isin 2009), whose voices fail to be reached or to emerge. These are certainly migrants, especially those with an irregular status living in precarious housing conditions (cohabiting in overcrowded conditions, squatting, etc.), but also and more broadly people living in precarious situations (such as the above-mentioned categories of elderly and fragile people, e.g. people with disabilities or psychiatric pathologies, etc.), regardless of their national background. The discourse on citizenship here is deeply connected to the discourse on the 'right to research' (Appadurai 2004), which is the right to access tools to increase knowledge capital, which is essential to fostering local claims and encouraging participation—at different levels—in policy arenas.

In the disadvantaged areas of our cities, people experience lower levels of access to opportunities and information compared to people in other parts of the city. Hence, many do not even know that they can aspire to be engaged in projects, activities, training, etc. which would potentially promote their personal development and potentiate their know-how. In this regard, Appadurai highlights the need to look at rights not only as a formal status but also by measuring the ability of different subjectivities to effectively put them into practice. In this sense, the development of competencies is intended to reinforce and stimulate the participation in society of people who have always been excluded from it and from the possibility to contribute to transforming their own environment. In this perspective, the author identifies the potential of protest as the main tool that can enable subordinate groups to 'critically participate' in public life and, consequently, in institutional life.

Working with and on local skills then represents an integrated approach oriented towards highlighting, reinforcing, or building individual competencies, not only for the sake of personal paths but also in terms of reinforcing each person's contribution to community life. In this regard, working on competencies not only acts at an individual level but also functions as a vehicle to promote collective reflections, directly involving local people and their personal expertise of experiences and knowledge of larger renewal processes. We depart here from the framework of urban regeneration defined as a social process and a policy establishing a relationship between space and society. Urban regeneration is a 'cultural approach', as Appadurai (2013) points out, as capacities are never isolated and are always part of a local set of means and ends, values and strategies, experiences and hypotheses put to the test. Therefore, a deep empowerment process should consider the relational character of urban regeneration, in order to be complete and more focused on a comprehensive development of the self.

5.3 Urban Living Lab: An Everyday Cultural and Plural Spatial Approach

Applying the lens of urban regeneration to the above considerations means to embrace a perspective focused on locally expressed (or unexpressed) individual and social skills, especially those of fragile residents, as an active component of the revitalization of large-scale social-housing estates. In the field of urban regeneration, as explained above, employing the ULL approach has been seen as promising in relation to the possibility to enlarge urban governance and effectively provide social actors access to participatory processes (Nesti 2018; Steen and van Bueren 2017). Indeed, ULLs seem to be able to develop and reinforce people's ability to reflect, imagine, aspire, and take action for the transformation of their life context, becoming real agents of change (Bylund et al. 2020).

But under which conditions can this activation take place, especially in poor and deprived contexts such as large-scale social-housing estates? Here again, relying on the ULL literature, it can be stated that it is crucial to recognize residents not only as beneficiaries of the interventions but also and primarily as coproducers, looking not so much to the response to needs but to the search for people's key aspects for engagement. In this perspective, ULLs become devices of inclusion if they are able to strengthen and improve local development and promote activities that are everyday, cultural, and plural, as will be articulated below.

First, activities should be everyday, because they directly connect to the practices of dwelling and living: day by day, they can mature within the direct experience of a space, a service, or a set of relationships. From this point of view, acknowledging the aspirations of residents means co-planning micro-interventions that are widespread with small targets, which would significantly improve the quality of everyday life. For example, the reactivation of small commercial spaces, the daily use of large green areas, the establishment of new recreational services (such as cultural activities) and services for families (such as after school activities), or the recycling of bulky waste abandoned along the streets of the neighbourhoods. Codesigning these opportunities to support the needs and wants a system of residents could be a starting point for initiating more structured solutions that address daily living.

Second, activities should be cultural, because they refer to the imagination that people cultivate with respect to their possibilities to change their life context; and the possibility the context itself allows them to be changed (Lazzarino 2017). In this sense, urban regeneration can work on the recognition of one's own and others' aspirations. Working on aspirations also means translating desires into instances of action for oneself and for the community. The role of imagination is fundamental, since it enables people to think differently and critically about common and recurrent situations. It would mean improving educational programmes and peer-to-peer initiatives, valorizing training events, and bringing out local leaders and representatives of the community. The training and strengthening of skills become two areas of intervention directly connected to the improvement of daily life, as they invest in social roles and professional paths, affecting both the 'from inside' position of residents and their ability to generate income, gain work experience, and build opportunities for socio-economic redemption.

Finally, activities should be plural: they must be diversified because they are a sign of belonging to different social groups and refer to very different life and work projects. Different groups of residents, with different characteristics, can be represented and identify themselves with the neighbourhood if activities are plural; emancipation paths must in fact be based on accompaniment and support, respecting the singularity of profiles and life paths. This implies a strong relational attitude, linked to individuals or small groups in a larger community.

In addition to these daily, cultural, and plural dimensions, the core element of urban regeneration promoted through the ULL method is the transformation of space, as a keystone capable of bringing together the everyday trajectories, cultural characteristics, visions and social complexity that distinguish social-housing estates. Hence, the transformation considered here is not so much the object of urban regeneration but is indeed its subject, as it assumes the role of 'collector' of forces and intentions, capable of bringing together the neighbourhoods' latent and diversified resources linked to a social and cultural challenge. Space is also a device for developing social self-representation, often humiliated and flattened within marginal contexts, through processes capable of preserving and reactivating small or medium portions of space as common goods. Underused buildings, abandoned public spaces, back courtyards and degraded green areas then become devices with which to aggregate and enable people and communities. The very typical and problematic characteristic of the social-housing estate's landscape—the widespread presence of run-down spaces —becomes a local resource, fostering the post-growth paradigm that aims to define development outside of growth, without building new space and consuming more resources (Cox 2017; Rydin 2013). Indeed, the challenge is to imagine forms of 'return' (of resources, spaces, centrality, and importance) to the territories that more than others have suffered the negative effects of dominant development models.

ULLs therefore consolidate 'social infrastructures of everyday life', in which people are enabled to take action driven by a desire for recognition and equity. The transformation refers to a multiplicity of uses, activities, and practices and is aimed towards the strengthening of open and welcoming spaces, which can speak different languages to a variety of audiences and protagonists. They focus on articulated dimensions of change pertaining to different fields: mechanisms for activating people as individuals or as groups; forms of consolidation of networks, roles, and powers within urban and local communities; and new shapes and functions of spaces and the landscape. We are referring to a circular relationship that puts in tension two poles of a possible change: on the one hand, how people and subjects place themselves in a new political and community dimension and, on the other hand, how places welcome and shape these renewed social aggregations through hubs and networks.

5.4 Frugality as a Tool for Participatory Urban Regeneration: The Case of the SoHoLab Pilot Projects

As noticed in the introduction, the ULL approach associated with urban regeneration has rarely been applied to very marginalized neighbourhoods. Therefore, the involvement of extremely fragile populations has hardly been experimented with, despite the large diffusion of the Living Lab theories and practices. For this reason, in the framework of the SoHoLab project, a set of activities was dedicated to the development of small pilot projects, intended to stimulate and activate the empowerment of local actors and residents through their direct involvement in colearning, codesign, and cocreation. Indeed, a socially oriented ULL—intended as. process-centred rather than product-centred (Franz 2015)—appears to be a space of 'empowerment through innovation' and 'innovation through empowerment'. Such a Living Lab is methodologically innovative, for it involves often excluded and marginalized populations in the design of projects with the aim of reinforcing their voice and competencies. At the same time, however, the ULL method is innovative in the development of more effective solutions to certain problems, for these are codesigned together with their own future users.

Indeed, because of its own nature, the ULL constitutes a tool which represents *par excellence* the possibility of testing innovation on a small (even micro-) scale, on the one hand enhancing the abilities and know-how of the people directly involved and on the other giving rise to solutions that can possibly be scaled up and transferred to other contexts or a wider scale. Regarding this aspect, Concilio (2016, p. 12) refers to the concept of frugality elaborated by Molinari:

> [...] experimented solutions use small amounts of resources and are frugal [...] from two different points of view. To begin with, they are developed with resources available in the specific problem contexts and do not require relevant additional economic or physical resources (citizens are more prone to mixing resources than professional designers). Secondly, they are developed and tested in spaces of proximity and localities. They are situated and consequently frugal in dimension and do not require large investments. This frugality adds to these solutions being reversible and effective in urban environments.

In this regard, ULLs—meaning places/processes in which to 'test' local solutions (pilot projects)—represent powerful devices in large-scale social-housing estates in at least two respects: first, in spatial contexts with a structural lack of public investment, small and 'frugal' projects targeting local problems have to absolve the function of 'testing' possible responses which—when successful—could encourage further public (or even private) investments and support a scenario of change, while offering concrete 'solutions' (even if restrained in scale) to residents. Second, the small scale and profoundly flexible methodology, characterized by the development of codesign and cocreation (as will be described in the next paragraph), offers the possibility of triggering and encouraging more effective local involvement, especially when dealing with populations that are structurally excluded from traditionally promoted forms of participation. In other words, as user-centred open innovation ecosystems (Pallot 2009), ULLs are 'sensitive' to the specific characteristics of the different users, which in the case of urban regeneration processes are the residents of a certain neighbourhood. As previously underlined, this specific feature of the methodological approach here considered is particularly relevant in large-scale social-housing estates, since it seems more effective when dealing with profiles characterized by multiple and layered identities (Association for Women's Rights in Development [AWID] 2004) and fragilities. On the one hand, working with competencies in the regeneration of large-scale social-housing estates is, indeed, potentiating and valorizing existing local knowledge, mostly referring to the valuable 'dwelling competence' employed by residents (La Cecla 1993); on the other hand, it means reinforcing and strengthening personal competencies and skills that are not yet fully expressed or recognized as vehicles to promote empowerment and stimulate local regeneration in terms of socio-economic improvement for residents and communities.

Within the SoHoLab framework, as already mentioned, several pilot projects were developed by the two research units in charge of implementing existing (Milan) or new (Brussels) ULLs. Both in the Milanese (San Siro neighbourhood) and in the Belgian (Peterbos neighbourhood) case, two of the three pilot projects activated in each neighbourhood directly involved residents, in the design phase as well as the implementation phase.[7] In the San Siro case, these pilots were respectively directed to the regeneration of a very compromised public space (Green Living Lab) and to the mapping/strengthening of the skills/competencies of a group of women of foreign origin[8] living in the neighbourhood ('Ghe pensi mi', which means 'I will take care of it' in Milanese dialect). In the Peterbos case, the pilots which directly involved residents dealt with the production of a new multimedia narrative of the neighbourhood (Digital Storytelling) and with the activation of a repair/upcycling lab (Brico Recontre) managed by residents. It is worth noticing that, on the one hand, the Green Living Lab and the Digital Storytelling pilot projects were more focused on the valorization of already existing 'nonexpert' local knowledge (colearning), while on the other hand, 'Ghe pensi mi' and Brico Recontre were particularly concerned with the emergence of new or not fully expressed competencies through an empowerment process (cocreation). Even so, these different phases are deeply linked to one another: as in the case of action-research approaches, with which the Living Lab methodology shares certain characteristics (Cognetti and Maranghi 2019), the pilot project tool (action) is crucial for gathering actionable knowledge (Argyris 1995) on a certain issue, advancing its understanding (research) in order to produce more effective solutions (action again).

[7] In both cases, the third pilot was dedicated to the development of a local network of third-sector organizations.

[8] A population which is particularly relevant in San Siro neighbourhood, not only numerically (half of the neighbourhood's population is of foreign origin, more than double the city average) but also in terms of being very active yet unheard by local and city-level institutions and also often 'socially invisible'; see Castelnuovo and Maranghi (2018).

5.5 Adapting the Living Lab Methodology: The Prefix 'Co' as an Empowerment Tool

Following the example of Franz (2015), who proposes a conceptual translation from technologically centred into socially centred Living Labs, we focus here on understanding which dimensions and tools proposed by the ULL methodology seem to be more relevant and effective in its 'socially oriented' definition applied to deprived areas of our cities. First of all, as illustrated in the above paragraph, it could be stated that pilot projects function, in these particular areas, as designed boundary objects (Concilio 2016) capable of activating different social worlds. Especially in large-scale social-housing estates, where many different cultures and backgrounds meet—many of which struggle to be fully expressed—these objects should be concrete (a space, a problem) rather than abstract (a concept) and clearly defined in space and time. Therefore, they usually function in a 'frugal' dimension and are capable of activating small groups of people at a time, who share a particular and concrete interest.

Through the development of such pilots, social ULLs becomes part of an empowering process which tackles the exclusion and disadvantage in at least two respects: on the one hand, it stimulates a sort of 'everyday' and individual activation; on the other, it impacts the empowerment of citizens in a wider scenario of urban change and regeneration. Indeed, through the direct involvement of their knowledge and know-how, people are stimulated to recognize certain abilities in themselves and see these abilities as tools to trigger the possibility to improve their living conditions in a certain area (typically work related). Moreover, as these pilots are part of an integrated and wider vision of regeneration for the area, they also stimulate residents' capacity to 'be citizens' and, for instance, actively participate in the redevelopment of local spaces and economies or the advocacy of a certain issue at the institutional level, with effects on the community as a whole. As an example, referring to the above-described SoHoLab pilot project, this was the case of the 'Ghe pensi mi' project, through which each woman of the group involved became more aware of her needs and competencies related to the job sector and, at the same time, became an active part of a wider appeal to institutions and local organizations to consider their struggle to be involved in the labour market, with an effect on the development of new policies and projects taking into account this perspective.

To be effective in these regards, in this typology of the ULL approach, particular attention should be paid to the relevance of the prefix 'co', since ULL deals with places and people with restricted access to decision-making processes. Here, indeed, specific attention should be paid to effectively 'include' local knowledge and voices. Although in the existing literature, it is not so widely described, when coping with marginalized contexts, codesign and cocreation should be based on a solid coresearch phase: it is, indeed, the phase which opens up the process and in which empowerment begins to take place through the development of colearning and the valorization of mutual learning. One of the most powerful characteristics of the Living Lab methodology is, indeed, the fact that it takes all different types of knowledge that very different actors bring to the process (activist knowledge,[9] usage knowledge,[10] professional knowledge,[11] and so on, see Nez, quoted by Lehmann et al. 2015) into profound consideration. To make this possible, the Living Lab should configure itself as a 'space of encounter' in the areas in which these different pieces and types of knowledge actually 'meet', according to Franz, taking place where targeted residents already exist and interact. In this regard, we recall the importance for the social ULL to be 'situated', so as to cultivate *geographical embeddedness* (Voytenko et al. 2016) as an element which fosters and nourishes the process of colearning through *contingency* (Karvonen and Van Heur 2014).

5.6 Conclusions

As underlined in the contributions, ULLs are capable of mobilizing material and immaterial elements simultaneously: the dimension of social bonds and the ways in which a person becomes a resource within an inclusive and proactive experience, gaining a role in micro-local community dynamics and the dimension of the space that becomes a 'thickener' of different tasks, functions, and meanings, sometimes even feeding new forms of local economy.

In these regards, the codesign and cocreation of small and incremental pilot projects, which engage institutions, local organizations, and residents, is indeed a powerful tool to rebuild a learning-friendly context (usually previously compromised in large-scale social-housing estates by long-lasting conflicts), fertile for acquiring the capacity of working together and generating new forms of governance. Moving, as suggested, between spaces and communities and working with incremental dimensions of change has proven to be effective but it takes time; it also keeps together what is already present in the territories in terms of resources or opportunities of a spatial and social nature and new trajectories of a contemporary city. We are referring to the interface between the actions of ordinary people seeking to change the places of their daily life and urban regeneration, between balanced planned intervention

[9] 'Activist knowledge […] is based on formal and informal knowledge of the administrative and political processes. From an individual perspective, it involves knowledge transfer and know-how acquired through one associative membership and belonging to informal networks. From a collective perspective, this type of knowledge is associated with the level of proximity and interactions between citizen collectives and the administrative institutions' (Lehamann et al. 2015, p. 1095).

[10] 'Usage knowledge is derived from a refined local knowing of citizens about a particular territory, which comes from repeated usages of product, infrastructures or services over time. This type of knowledge is usually externalized through stories and testimonies, revealing the particularisms of a given territory as well as usages conflicts over it. Collectively, this kind of knowledge will be formalized through public debates and the expression of "common sense"' (ibid.).

[11] 'Professional knowledge is derived from the technical skills of particular stakeholders in the LL. While experts often generated this type of knowledge, it can also emerge from layperson whether from their belonging to a particular group, formal or informal (i.e. makers) or from professional skills acquisition in the LL itself along the road. This knowledge can also emerge from the interactions between stakeholders within the LL leading to collective professional knowledge' (ibid.).

and self-organized practices. Urban regeneration is a long-term process that must be anchored to the genius loci of a place and should focus on urban software (the community) and not only on its hardware (urban space) (Ostanel 2016).

In this respect, ULLs able to stimulate empowerment are somehow an apparatus for 'insurgent regeneration' (De Carli and Frediani 2016; Holston 1998; Miraftab 2009; Paba 2002). They function as 'radical devices' capable of wisely putting in place a delicate balance between the present and future (Cancellieri 2019). On the one hand, the regeneration project, by its very nature, is necessarily strongly oriented towards change and therefore towards a projection forward with respect to possibilities. On the other hand, it must be able to intercept and support what the territory already expresses in everyday life and within the experience of places, often starting from existing projects and ongoing processes. This means creating a future projection by anchoring itself to current conditions, in search of a difficult balance aimed at introducing new elements that support change as well. In this way, the ULL should function both as a long-term framework and as an everyday activator, flexible enough to adapt to unexpected events.

Hence, regeneration becomes a process and is no longer a preventive intention but an idea that is substantiated by being implemented, configuring itself as an open and evolving form within which the Living Lab becomes a tool that effectively supports the contribution of different stakeholders and social actors to the process. Within this delicate balance, the issue of the structural lack of basic social policy (housing, education, labour market, etc.) remains open; in other words, the development of such a process should not hide the fact that, in some cases, democracy and participation must be pursued with a surplus of 'very social policies' (Tosi 2017), tuning it to the demand of poorest or which even the most innovative projects alone are not enough. In this sense, however, ULLs seem to be promising, as long as they are able to empower communities and individuals to activate themselves not only in producing innovation but also in claiming their rights.

References

Appadurai A (2004) The capacity to aspire. In: Held D, Moore HL (eds) Cultural politics in a global age. Oneworld Publications, pp 29–35

Appadurai A (2013) The future as cultural fact: essays on the global condition. Verso Books

Argyris C (1995) Action science and organizational learning. J Manag Psychol 10(6):20–26

Association for Women's Rights in Development (2004) Intersectionality: a tool for gender and economic justice. Women's Right and Economic Change 9:1–8

Bylund J, Riegler J, Wrangsten C (2020) Are urban living labs the new normal in co-creating places? In: Smaniotto C, Menezes M, Mačiulienė M, Romeris M, Goličnik Marušić B (eds) C3Places, culture & territory, vol 4. Lusófona University Press, pp 7–21

Cancellieri A (2019) Attrezzi per una rigenerazione urbana radicale. In: Balbo M, Cancellieri A, Ostanel E, Rubini L (eds) Spazi in cerca di attori/attori in cerca di spazi. La rigenerazione urbana alla prova dell'innovazione sociale. Master U-Rise, in collaboration with cheFare, Tipolitografia Pavan

Castelnuovo I, Maranghi E (2018) Forme dell'abitare femminile migrante. Il quartiere come occasione di capacitazione? In: Cognetti F, Padovani L (eds) Perché (ancora) i quartieri pubblici. Un laboratorio di politiche per la casa. Franco Angeli, pp 97–114
Cognetti F, Maranghi E (2019) Adapting the Urban Living Lab approach to marginal contexts and urban regeneration: the case of Mapping San Siro Lab. In: Proceedings of open living lab days conference 2019. Co-creating innovation: scaling-up from local to global. European Network of Living Labs, pp 200–206
Concilio G (2016) Urban living labs: opportunities in and for planning. In: Concilio G, Rizzo F (eds) Rethinking the interplay between design and planning. Springer, pp 21–40
Cox KR (2017) Revisiting the city as a growth machine. Camb J Reg Econ Soc 10(3):391–405
De Carli BA, Frediani AA (2016) Insurgent regeneration: spatial practices of citizenship in the rehabilitation of inner city São Paulo. Geohumanities 2(2):331–353
Franz Y (2014) Chances and challenges for social urban living labs in urban research. In: Proceedings of open living lab days conference. European Network of Living Labs, pp 105–114
Franz Y (2015) Designing social living labs in urban research. Info 17(4):53–66
Freire P (1970) Pedagogy of the oppressed. Herder & Herder
Hawlik R, Berger K (2021) Era-net cofund. Smart urban futures, Synthesis report, Urban Europe. https://jpi-urbaneurope.eu/wp-content/uploads/2021/09/ENSUF-Synthesis-Report-AIT-FINAL-211103.pdf
Holston J (1998) Spaces of insurgent citizenship. In: Sandercock L (ed) Making the invisible visible. A multicultural planning history. California University Press, pp 37–56
Isin EF (2009) Citizenship in flux: the figure of the activist citizen. Subjectivity 29:367–388. https://doi.org/10.1057/sub.2009.25
Karvonen A, Van Heur B (2014) Urban laboratories. Experiments in reworking cities. Int J Urban Reg Res 38(2):379–392
La Cecla F (1993) Mente locale. Per un'antropologia dell'abitare. Eleuthera
Lazzarino E (2017) Mercato Lorenteggio. Un approccio culturale alla rigenerazione urbana. In Andorlini C, Bizzarri L, Lorusso L (eds) Leggere la rigenerazione urbana. Storie da 'dentro' le esperienze. Pacini Editore, pp 57–64
Lehmann V, Frangioni M, Dubé P (2015) Living Lab as knowledge system: an actual approach for managing urban service projects? J Knowl Manag 19(5):1087–1107
Malmberg K, Vaittinen I, Evans P, Schuurman D, Ståhlbröst A, Vervoort K (2017) Living Lab methodology handbook. Zenodo. https://zenodo.org/record/1146321#.YiUAG9_SJpQ
Miraftab F (2009) Insurgent planning: situating radical planning in the global south. Plan Theory 8(1):32–50
Nesti G (2018) Co-production for innovation: the urban living lab experience. Policy Soc 37(3):310–325
Nussbaum M (2002) Giustizia sociale e dignità umana. Da individui a persone. Il Mulino
Nussbaum M (2011) Creating capabilities. The human development approach. Harvard University Press
Ostanel E (2016) Urban regeneration and social innovation: The role of community-based organisations in the railway station area in Padua, Italy. J Urban Regener Renew 11(1):79–91
Paba G (ed) (2002) Insurgent city: Racconti e geografie di un'altra Firenze. Mediaprint
Pallot M (2009) Engaging users into research and innovation: the living lab approach as a user centred open innovation ecosystem. Webergence Blog. https://web.archive.org/web/20120509081658/, http://www.cwe-projects.eu/pub/bscw.cgi/1760838?id=715404_1760838
Rittel HWJ, Webber MM (1973) Dilemmas in a general theory of planning. Policy Sci 4:155–169
Rizzo A, Ståhlbröst A, Habibipour A (2021) Transformative thinking and urban living labs in planning practice: a critical review and ongoing case studies in Europe. Eur Plan Stud 29(10):1739–1757
Rydin Y (2013) The future of planning: beyond growth dependence. Policy Press
Sen AK (1985) Commodities and capabilities. North Holland

Sen AK (1990) Development as capability expansion. In: Griffin K, Knight J (eds) Human development and the international development strategy for the 1990s. Macmillan, pp 41–58
Sen AK, Nussbaum M (1993) The quality of life. Clarendon Press
Steen K, van Bueren E (2017) The defining characteristics of Urban living labs. Technol Innov Manag Rev 7(7):21–33
Tosi A (2017) Le case dei poveri. E' ancora possibile pensare un welfare abitativo? Mimesis
Voytenko Y, McCormick K, Evans J, Schwila G (2016) Urban living labs for sustainability and low carbon cities in Europe: towards a research agenda. J Clean Prod 123(1):45–54

Part II
Positioning Research(ers) in Large-Scale Social Housing Estates

Chapter 6
Beyond the Presence: Dwelling *with* People and *with* Their Places

Ferdinando Fava

Abstract The author intends to indicate some epistemological and political nodes of 'being there' at the centre of ULLs, in different forms as implied by the SoHoLab project. At the root of the idea that urban sites can provide an arena of learning within which the co-creation of innovation can be pursued among research organisations, public institutions, the private sector and community actors, lies the possibility of establishing meaningful relationships as a medium to know these sites, construct social design, implement and govern local and national housing policies. In the light of the modus operandi of anthropological field research, on another way to 'being there', the author shows how 'these meaningful social bonds' to be epistemologically and politically relevant need to be coupled with a strong critical reflexivity able to deconstruct continuously the discursivities (of policies, of disciplinary as common and mainstreaming narratives) and practices of the ULL itself. A cognitive strabismus has to be developed to catch these place-based laboratories and contexts dependents, to make them 'up close', apprehend 'from inside' and 'from below'. Analysis situ and analysis in situ are not disjointed: the third space of knowledge construction allows to join them and recognise the logics that govern these social bonds.

Keywords Immersion · Knowledge · Ethnography

F. Fava (✉)
Università degli Studi di Padova, Padova, Italy
e-mail: ferdinando.fava@unipd.it

Laboratoire Architecture Anthropologie, UMR LAVUE 7218 CNRS, ENSA Paris La Villette, Paris, France

N. Aernouts et al. (eds.), *Urban Living Lab for Local Regeneration*, The Urban Book Series, https://doi.org/10.1007/978-3-031-19748-2_6

6.1 Introduction[1]

The European Network of Living Labs (ENoLL), an umbrella organisation for living labs around the world, defines living labs as 'user-centred open innovation ecosystems based on a systematic user co-creation approach, integrating research and innovation processes in real-life communities and settings'.[2] This definition condenses a broad range of domains and organisations that make living labs the main research facility of action strategies (sometimes carrying messianic expectations) for the innovation of the millennium. The literature concerning this topic has already become noteworthy (to quote only a few: Chronéer et al. 2019; Hossain et al. 2019; Marvin et al. 2018) and relates to an array of key features that make these devices appealing to both public and private entrepreneurial partnerships. The body of work on living labs gathers heterogeneous elements: from *landscapes* to *real-life environments*, from *methodologies* to the inclusion of public and private *stakeholders*, and different business *models*, methods and approaches. Open innovation works here as both a technical and economic paradigm in which centre is the final user, involved from the very beginning as cocreator and, at the same time, as beneficiary of the lab.

The *urban living labs* to which some of the chapters of this book will refer definitely fall within the frame of the *mainstream* definition quoted above, especially taking into account *tenets* like the integration of research, cocreation and innovative planning in the users' real-life contexts. Nonetheless, they appear to be eccentric variations, due to the epistemological, social and political nature of the 'innovation' that they want to promote. Situated in urban contexts, particularly in the so-called marginal residential areas, i.e. *public housing projects* that became urban 'black holes', these living labs are driven by the purpose of urban regeneration.[3] In their conception, urban regeneration is neither a new consumer product to be launched in the market (Nesti 2018) nor is an outcome of social technocratic engineering or much less an architectural *commodity*. The practice of urban regeneration is here a relational process that assumes the epistemological and political recognition[4] of the social bonds that entail dwelling in a place. This acknowledgement allows for the cooperative co-creation of initiatives aiming at transforming that very place together

[1] This text summarises the keynote from a seminar held online on 13 October 2021 as a part of SoHoLab project. For more information, see the Introduction of this book.

[2] https://enoll.org/about-us.

[3] 'The complexity and uncertainty of the contemporary city and urban areas such as large-scale social estates require new readings, interpretations and analytical lenses. Regeneration cannot be thought remotely but should be rooted in place'. From the flyer of the SoHoLab seminar on 13 October 2021.

[4] Recognition is of course a concept that can be applied to different semantic orders. It refers to intersubjectivity (Ricœur 2004) and to political dimensions as well, concerning epistemology (Honneth 2005) and justice (Young 1997). The reasons for this polysemy are epistemological: the possibility of recognising the actors as carriers of a critical knowledge not to be devalued due to their ascribed social identity or locationing or for the lack of interpretative categories (Fricker 2007). For an introduction to recognition as a political concept, see van der Brien and Owen (2012) and Thompson (2006).

with the actors living there, and it does become the ultimate goal of urban transformation itself. In short, social recognition is the matrix, effect, and style of the living labs' way of proceeding.

In these pages, I will bring the reader's attention to what I think is the epistemological kernel of a key common feature of mainstream living labs and socially oriented *urban living labs* discussed in this book, that is, the explorative and cognitive research on the contexts in which they are located, their 'ecosystems'—the definition quoted at the beginning borrows the term from Population Biology, therefore overshadowing power relationships and the neoliberal environment where living labs take place. I will put into dialogue the living labs' cognitive endeavour with the most familiar practice for me: long-term anthropological field research, based on interindividual interaction and the *ur-practice* of conversation. In this respect, I propose a reflection on the use of ethnography and all other forms of direct and long-running encounter by *living labs* in order to comprehend the social universe of urban dwellers, starting from the permanent settlement in these residential areas,[5] whether it is the collective space of a lab or the living/workshop of a single person. Here, long-lasting ethnography appears as a research strategy in accordance with the living labs' reasons for being; the immersion in everyday life and social practices is an integral part of the process of transformative planning. That is of course a complex task, centrifugal in relation to the disciplinary safety of urban planning and architectural design, and also unpredictable, for such a project of immersion always entails a new path to trace. Indeed, for every single case of analysis, we just have to invent how this 'immersion' takes place and, hence, how it can give rise to a methodical observation of people, places and the process of planning itself. Ultimately, the manner in which the knowledge thus produced is used in the planning process cannot be predicted: this necessary and founding 'local knowledge' needs to be processually integrated into methodological-disciplinary apparatuses that are not very flexible, refractory or even impenetrable to 'contaminations' from 'irregular' knowledge.

In the following pages, after sketching the social ontology[6] of the SoHoLab project as well as the practice of anthropological research as I do it, I will outline the trope of 'Malinowski's tent', a figure of 'immersion' characterising anthropological research in the field, and reread it along with other figures of immersion promoted by the living labs in this book or from which they take their cue. It will become clear how settling into a place is not sufficient for knowing it. Extra critical reflection is necessary so that the adoption of these figures/tropes fulfils the promises coveted by living labs. Ultimately, I will conclude by illustrating how the innovative experience

[5] 'This practice of "situating" in space and place can be introduced before the planning process, undertaken simultaneously or developed throughout the development process, preferably constantly contaminating the practice', from the flyer of the SoHoLab seminar on 13 October 2021.

[6] I refer to Keith Sawyer's reflection on the concept of emergency (Sawyer 2005). I refer also to Marguerite Archer's works, especially those from *Structure, agency and internal conversation* (Archer 2005) for her contribution to 'individual agency' as a project mediating constrictions and facilitations of 'causal powers' within social and cultural structures.

of the SoHoLab projects can lead to rethinking the act of dwelling by turning both spatially and epistemologically to the places and those who live them so that urban regeneration can be rooted in the places since it is part of them.

6.2 In the Beginning, Living Together and Figures of Immersion

According to linguists, 'living together' is an everyday, colloquial, almost banal saying, which seems to me evocative yet substantial to express what living in a place is and to define what places are 'made of'—basically, built space and human bonds. Architecture, buildings and streets have a meaning thanks to relations, in particular to the kind and quality of the bonds we establish through them, which these buildings and streets in turn iconise, index (in a Peircian sense), enable and relate to. We can say that in the beginning are the bonds. It is a category on which I confer a connotation that is firstly ontological and epistemological, with methodological derivations that I will illustrate, without denying the political, ethical and poetic harmonies coming with it. It is precisely this expression, social bond, that guides my rereading of the *modus operandi* with which some of the SoHoLab projects practise and interpret immersion in their 'doing', by putting it into 'action'.

By means of the interpretative *phronesis* of the SoHoLabs, the images of the immersion and the root—both behind the idea of a regeneration of the place pointed out, guided and coming from the inside—become courses of action. These two metaphors represent an alternative, critically and intelligibly performed, to *mainstream* top-down urban and social policies. Yet, in so doing, they open paths to be explored, new ways of proceeding which can always be readjusted. The immersion in social practices and the everyday life of places is necessary to think this regeneration from near, from the inside and from below—from the 'roots'—in a cooperative way. Of course, the imaginary of immersion is itself ambiguous. We anthropologists are responsible for the idea that 'entering' a culture means 'immersing into it'. This is misleading if we push that theory to the far end of its literal meaning. Immersion would imply the idea of an environmental passage, a radical change in status. Now, it is clearly not like that. Stressing the differences is part of the strategy to stage the overcoming of the differences. Social changes on a global scale—and before them, a critical grasp of the act of creating anthropological knowledge—blew up the ideology that *people*, *place* and *culture* are welded together, of which area studies were a direct consequence. That ideology on one hand led to the idea of 'entering' an alien culture by gaining the trust of its members, the informants. On the other hand, once anthropologists started exploring urban areas, they were still looking for the village, the microsociety, selecting specific neighbourhoods, or those elements within the neighbourhoods that seemed to confirm the confinement of microsocial in microlocal. National belonging, 'ethnicity' and social class were deemed as those features allowing one to retrace the place for a common belonging in the heterogeneous city. Actually, once we are in the field, whether it is urban or rural areas, we do not interact with cultures—be they national, ethnic or of poverty. Rather, we get to meet living and concrete people and we interact with their doing, their gestures. We enter a living relationship with their space–time. We meet people who belong to groups with internal hierarchies, borders and social universes: not static, but rather caught in their becoming. We are necessarily in the places and their historical temporality, made of conflicts and ruptures, of the bonds that form them and that are moulded so. This immersion—as long as we keep on using this image with this warning in mind—is the device allowing SoHoLabs to interact with the residents in a worthwhile way during the whole process of a planning which otherwise, it has to be highlighted, has often been devised from afar, from the outside, by a few experts.[7]

The artist residence (Aernouts et al. 2020, pp. 147–152), the *permanence architecturale* (Hallauer 2015), and the living lab are all figures, each of which has its own constellation of variations in a hypothetical taxonomy, of staying in the places while being equipped with critical thinking. This falls, on one hand, within the practical and cognitive value horizon of art, architecture and urban and landscape planning and, on the other hand, within the relations and bonds that these actors establish with the people living in these places. These are ways of being in the neighbourhoods that, once they are put into action, call into question the epistemology and disciplinary *mainstreaming* practices, which made living in those places a specific object of analysis and planned action. By acting like this, the usual devices on which these apparatuses rely are yet unbalanced and questioned in a reflective way. The epistemological devices tell apart and classify theoretical and practical knowledge through hierarchies of value, along with dividing expert knowledge from common sense. Yet, they often mistake with the theoretical projection of a transparent conscience of oneself on others' practices with the sense 'created' from the doing, which instead stems from the needs of everyday life. The political devices handle the distribution of powers between public deciders, stakeholders and residents, activating formal decisional processes and authorising parodies of participation. The mediatic devices fuel dominant representations proposed as true and in so doing distribute the values of normality in and out of these places or, on the contrary, promote different restitutions of them in the public sphere. In this 'being' in places and weaving these bonds, an awareness—one of the results of the endeavour of this immersion—is reached of the kernel of the values guiding judgements amidst the life forms that the living labs mean to understand. At the same time, this is something I want to stress, one becomes aware of the values and evaluations of the gesture of those who mean to understand them. In other words, in these figures of the immersion in places, the epistemological, political and ethical fragility of thinking of the others' dwelling without them surfaces. Thus, the result is not only a gap between a *savant* representation of dwelling and its social, concrete and situated factuality but also between the symbolic order of the residents and the symbolic orders of disciplinary fields and

[7] We continue to have the idea that an expert and disciplinary practice when it comes into contact with life is contaminated by it, such as COVID-19.

urban policies which rule and maintain this exclusion. Paradoxically, this separation is also reproduced in the solution that *prima facie* appears to solve it: the settlement *tout court* in the place as sufficient per se to understand the dwelling of others. It is a figure of a romantic and innocent *going resident*, an urban avatar of the *going native* of anthropological tradition. This solution presupposes the exclusion and keeps it unresolved, like a haunting ghost in the attempt to go beyond it, unless it is not deconstructed and traced back to the epistemological and political conditions that make it possible. To 'immerse' into the social customs of the places in order to grasp their social universe *from the inside* requires critical reflection, of course not self-centred, which resets this simple 'being in' as a more structured and complex 'being with'. The Scilla of the objectifying external gaze on dwelling and the Cariddi of a naive and transparent identification to the place and its dwelling frame the 'space' of the epistemological and political placement of this immersion. The ascetic exercise of an epistemological vigilance and faithfulness to the methodological disposition warn us not to choose these two options, despite being appealing comfort zones. Thus, via *negationis*, the SoHoLab and its actors are placed *in between*, within a threshold space that is the space of interaction.

The presupposition, often not made explicit, that founds living labs understood in this way is a social ontology of places which considers social bonds—the ones residents establish between them and with their built spaces—as both the matrix and the effect of dwelling. It is an ontology needing their recognition anyway. Upon closer look, it also represents one of the necessary conditions to practise anthropological fieldwork, as well as a compass for orientating methodological reflection. Indeed, anthropological research is based upon the attention to the microsocial level of interaction, which can be justified only insofar as such an interaction, through which bonds are established and kept, is thought to have a relative autonomy where we the actors elaborate unpredictable significant realities. The bonds that we weave doing research as much as those we aim to understand, everyday bonds always in the making, cannot be boiled down to global mechanisms of structural processes nor to stigmatising mediatic stereotypes. To put it in a way related more to contemporary social theory, the actor *agency* giving shape to these bonds eludes structural (spatial, economic, political and symbolic) constrictions, although it is bound to them. This agency creates an unexpected and unpredictable space of action which is therefore sensible. It is not possible to meet such a space of action outside the space–time of its happening; we must be within the dialogical interaction, in microsocial proximity. This modest yet real autonomy of invention allows for a field research based on the interindividual encounter, taking conversation as the practice where its knowledge originates. The microsocial level of this research encounter and of the bonds that the actors continuously establish between them in the field, therefore have an autonomy authorising the acknowledgement of the 'place' of the resident as the position from which they negotiate, resist and redefine the bonds that, on different scales, burden them and their space, marking their bodies and architectures. I wish to discuss further some points about this fieldwork bond, that makes the encounter an epistemologically relevant relationship.

6.3 Malinowski's Tent

I would like to add a fourth figure of immersion in social practices, which I would call 'Malinowski's tent' out of love for the discipline and henceforth 'tent'—the figure of the anthropologist's physical (as well as symbolic) living within the social universe (s)he means to analyse. (S)he centres this analysis not so much on consulting the archives or reports from others or even statistical *surveys* but mainly on his/her direct relationship with the subjects, who (s)he met where they produce their place and their story, which both affect their lives. Also, on one hand, 'being' in the anthropologist's field, as much as the other figures mentioned above, refers to the horizon made up of evaluations and representations of his/her discipline; on the other hand, it refers to his/her *modus operandi*, that is, the way of being in these bonds and in social practices that establish them. The interdisciplinary debate in the second half of the twentieth century was indeed marked by a deep questioning of epistemological and political presuppositions and how to perform this new theoretical approach in the field. Ethnography, in the version I am presenting here, is far more than a tool to gather information and autoptic observations: it is a path to follow, *meta-odon, a modus operandi*; a way to learn how the interlocutors establish their social bonds. This involves the anthropologist personally, since it is only through this bond that (s)he succeeds in establishing on site with his/her interlocutors that this understanding of a specific universe is made possible. The anthropological tradition—specifically, the one in which I learnt the profession—has something to say about how to make these interactions relevant from an epistemological point of view (Althabe 2001; Althabe and Hernandez 2004). The research relationship develops through a bond that is not present before the anthropologist gets to the place; it goes through the inscription of this relationship in its duration. The anthropologist does not act like the bird striking the hours in a cuckoo clock: (s)he does not get out for a moment and then gets back in and then disappears. (S)he weaves an emerging bond which indeed unfolds and stays that way over time by means of social acts of mutual recognition. It is a bond intentionally and clearly oriented towards knowledge, which stays so for the entire duration of the research because it does not lead the anthropologist and his/her interlocutors to hold a social, institutional or familiar role, in their different social universes. It is a bond oriented to knowledge, whose space is, therefore, as mentioned above, that of the threshold between insider and outsider, which makes understanding through dialogical construction possible by virtue of this in-betweenness (Fava 2017).

'Being there': as a bedrock of the figure of the tent, there is the founding gesture of 'departure'—which is usually a source of fascination—and 'a place' that has to be established as 'a relationship, a bond', or rather, a net of bonds in order to become a source of knowledge. In this surprising and defamiliarising net, questions but also restlessness, pain and joy arise. These two gestures of 'leaving' and going to a 'place' are at the root of what I call the anthropologist's topology, and contribute to the creation of emerging bonds, as I have already said. Not only are they emergent because they were not present previously—the encounter with the anthropologist always represents a rupture, a break in the everyday ordinary rhythm, but also because, despite being linked to the anthropologist's bonds, they are not circumscribed by them. These emerging bonds are *in-between* positions, other spaces of dialogue and observation, around which the everyday bonds and their vicissitudes, conflicts, and tensions can come to light, leading to a shared awareness. Through this emerging bond, it is, therefore, possible to learn of the group's other inner bonds, the bonds between individuals and institutions, the bonds with their built environment, the relationship between the contingency of interlocutors' present time and the temporal depth of socio-economic and spatial transformations on which their dwelling is projected. 'Immersing' into the social practices of the places means witnessing these bonds through 'Malinowski's tent', acknowledging and understanding them—I repeat—from the research bond in field.[8]

A further feature of the epistemological and methodological reach of the mutual recognition of bonds behind our social ontology is that we the social actors 'interpret' ourselves mutually when entering a relationship. We give a pragmatic meaning to actions, to the gestures of others as well as our and our interlocutors' social and spatial positions. This is an acted meaning, not explicit, which many times does not lead to a reflexive awareness, yet it rules our taking the floor in the dialogical exchange—or its denial—and the interaction, our 'what to do' and our 'where to go'. The matrix of this meaning lies in the personal horizon made up of the life story, values and the present lived by each one, where the effects of a shared social and economic contemporaneity arise, with which each one has to come to terms. In the way I conduct the anthropological fieldwork, this meaning plays an important role in achieving an understanding *from the inside* of the social universe of the people I meet. The emerging research bond, which suspends the usual bonds and opens the possibility of a knowledge relationship, originates from this operating attribution of meaning, anchored to the singularity of the interlocutors and of their socio-spatial situations. As is said, the observer is observed. Acknowledging this meaning in its spatiotemporal becoming—*who am I to my interlocutors?*—and its working in the research bond is a methodical goal in itself. It helps define the interpretative frame of the saying and doing happening in this bond, of which the anthropologist is the witness and coactor (in disciplinary terms: the ethnographic material that has been gathered, descriptions of observations and transcriptions of exchanges and interviews are communicative products) (Fava 2017).

The consequences of this prerequisite are different. Just two of them come to mind, which seem relevant to me to denote the immersion sought by the actors of the living labs. The first consequence is epistemological and concerns 'intimate' knowledge. It is thanks to this mediated acknowledgement of how the anthropologist is so engaged, that is, authorised to enter the bonds of his/her interlocutors, that their social universe can be understood in the present without reducing it to an objectivation from the outside, or trapping it in a persistent grid, or retyping it on a predictable

---

[8] In historiographic terms, I will point out that Malinowski understood his being in the village within the paradigm of positivism, where the relationship with the residents was understood as a mere tool for gathering information (Ellen 1984, pp. 48*ss*; Stocking 1983, pp. 7–120). The hermeneutic awareness of this relationship developed in research practice later (Geertz 1973, pp. 3–31).

and timeless representation or on a presentist narrative atomisation, concealing what presently matters to the interlocutors, *what is going on*, which refers to the history of individuals and places.

The second consequence is methodological: it is the decentring of listening and looking. The endeavour to acknowledge this implication leads to a necessary and continuous decentring of the self in the research process. Here, the self is the psychological, methodical and epistemological self of those who wish to immerse themselves. Listening to the interlocutor requires making room for his/her narration without worrying about handling his/her word. The gaze of the interlocutor crosses the gaze of the anthropologist, ceaselessly evading the latter, without ever becoming one of its objects, something visible, because that gaze observes and contains that one of the anthropologist. Forgoing the pursuit of his/her own research agenda anchors this agenda to the space-times of the interlocutors, tuning the research into the latter's becoming. The conceptual mediations that mean to account for this universe are therefore built from the inside of this interaction, which acknowledges not knowing as generator of a situated knowledge, always in transit.

I find these features to be decisive, so that knowing the places that the socially oriented urban living labs pursue matches the recognition of bonds, which is necessary for the cooperative elaboration of their possible transformation. Indeed, it is not just up to the anthropologist to wonder about the meaning that her/his interlocutors assign to immersive (non)cognitive practices and about the effects that these have upon them and their universe. In other words, it is also possible to wonder, as for the immersion figures mentioned, what meaning the residents assign to the artist residence, to the architectural permanence, and to the living lab to which the actors of this planning reach out? It is a reflexive question indeed, yet with a decentring reflexivity, which brings back to the residents, their troubles, and the process of creating the place—the latter, simultaneous to the research process, enables one to re-question the cooperative planning by redefining and reorienting it. The immersion, in order to be productive, needs that exchange of gaze mentioned above; the gaze of our interlocutors on our practices and our glancing back; that is, that critical reflexivity which is not a mirror of the status quo but rather an icon of the possible openings.

6.4 Which 'Immersion'?

In view of this rapid read-through of the practice of anthropological research, I wish to get back to the figures of immersion and reread them. If we line up *architectural permanence*, the artist's residence, and the living lab together with the tent closely, we can detect an accent from the same native language: the 'gesture of going' and the 'emerging bond' typical of the tent seem to create a common founding device which produces a knowledge relationship present in all of these figures, although with different purposes and outcomes. For sure, this device, as it has been remembered earlier, is inscribed in institutional traditions and processes which would strictly redefine—although not in an immutable way, as proven by SoHoLabs—the relationship between different actors involved: the architect, the artist, the *urban planner*, the anthropologist, and their interlocutors, in this case, the residents. This relationship is made up of a complex twine of words, gestures and possibilities. The everyday experience shows that these founding gestures, despite being so socially and disciplinarily built, are spread through options that always witness a margin of choice—a personal initiative—which lies at the origins of these figures themselves, in compliance with the social ontology of the places and consequently, defying disciplinary canons. These are the choice of the interlocutors; the choice of long durations; the choice of a decentring listening, which seizes in what is heard an instance of truth that exceeds the individual and the local. The choice of sifting through one's own discursive and material context, one's own agenda, the categories in use, especially those that seem obvious, clear and glaring. The choice of narrative strategies; the cautious choice of words to be used to write the reports; the choice of the intended audience; and the choice of taking into consideration the possible effects and possible political uses of what is told about these bonds—that is, these places—in the public sphere. The living labs' practice of everyday life proves once again how the personal word plays an unequal role and weight in the public sphere, especially for the residents of the *social housing estates* and their representation in the media.

This discerning, which is inseparably epistemological, ethical and political, opens the possibility of reconfiguring the structural asymmetries established in the relationship with residents from the outset. The founding gestures of the immersion figures, indeed, condition invisibly relationships and mutual positions. Thanks to these gestures and these bonds, some hold the positions of artist, architect, urban planner and anthropologist, whereas others are just residents of the neighbourhood. Acknowledging this means acknowledging how these bonds are marked by cleavages to which they are not reduced but of which they bear traces. The Heideggerian dwelling and Holderling's poetic dwelling are not deprived of relationships of power and exclusion, yet they bear their footprints as scars from wounds. Well, like Bourdieu, we would say that this discerning opens the possibility of decreasing—with patience and over time—the symbolic violence always looming over us, and making it ineffective by going through it (Fava 2021). Of course, the space of manoeuvre is modest, since it is personal, and limited, since it is real. It is the decentring effect once again: away from one's own universe of practices and one's own categorial horizon, from one's own epistemological narcissism, from one's own disciplinary, professional, even existential comfort zones. This leads to the genealogical and epistemological critique of one's own categories, that are thus 'mobilised' in the act of decentring itself. This also leads to re-establishing the mutual acknowledgement of the acting subjectivity through the differences and invisible diaphragms wherein these bonds are crystallised.

Social housing estates are often the recipient of social and urban interventions because the residents' personal initiative and voices, their capacity for self-representation, their critical subjectivity, the energy they spread out in order to live day by day and try to get away from the grips of material difficulties and external rejections—as the critical analysis of the SoHoLabs highlights—are massively denied. *Social housing estates* are neither obvious nor transparent.

The perspective opened by living labs is that of knowing from the inside, according to registers that are not limited to an instrumental presence in these places, as in the expression of a methodical antiseptic rationality. It is quite the contrary: it seems to me, from these figures of immersion, that being in these places is not just a physical presence aimed at gathering information, but rather an 'attempt to inhabit them', inhabiting a research relationship, endeavouring to stay there over time, alone or together, establishing bonds, placing at the centre not so much the architectures and their configurations as the concrete people and their relationships. Residents are unique singularities that carry shared wishes too, as we all do.

6.5 Conclusion

The planning experience of the living labs and the knowledge of *social housing estates* that they promote are abundant with significant repercussions that exceed the relevance of the local knowledge they build. In particular, as a conclusion to these brief reflections, I would like to stress their contribution when it comes to thinking about dwelling. I think it is also to be stressed that 'dwelling'—through this form of knowing by living the places via these figures of immersion—is no longer a mere object of disciplinary perspectives. It is also the subject *from* which to think. It is the passage from the epistemologies *of dwelling* in the places to an epistemology of the places *from dwelling*, from the inside, yet where the others' dwelling is not just the neutral object observed from a formalised discipline, tested negative for the contaminations of reality. The others' dwelling is also the subject, the experience lived and practised by residents and 'urban professionals' which establishes a cognitive and praxeological apprehension of dwelling as the critical experience of those who are the subjects of it. It is no longer possible to think of the dwelling of others without them. Their dwelling is no longer or not only a technical-engineering reduction, an abstract entity of reason, an exotic and romantic dream, a juridical-legal device or a construction of rules and policy tools. Without denying these acceptations, the dwelling that emerges from living labs is firstly a shared experience of understanding the bonds that make up a place. For it is in dwelling indeed that the differences in hierarchy, values, social status and gender arise and are made visible. The awareness of the epistemic breaking always involved in the endeavour to understand these bonds enables one to restore the conditions of an understanding suitable for the decentring of looking and listening, which leads to the social acknowledgement of the subjectivities we always are. It is not enough to install oneself in the *social housing estates* to conceive their urban regeneration. Living labs prove that it is necessary to understand the place of others: it takes a critical position, an intelligent grunt which is not ascribable to experts, theorists, deciders, nor the residents. It is a shared property, a variable-sum inner good, if it is apprehended (learnt), invented and carried out together. The stories of the living labs on the following pages show that it is possible, involving adventures certainly without guarantees, but promising and fruitful already in their realisation.

References

Aernouts N, Maranghi E, Ryckewaert M (eds) (2020) Towards a definition of socially oriented urban living labs. SoHoLab
Althabe G (2001) Pour Une Ethnologie Du Présent. Ethnologies (Québec) 23(2):11–23
Althabe G, Hernandez V (2004) Implication et réflexivité en anthropologie. J Des Anthropologues 98–99:15–36. https://doi.org/10.4000/jda.1633
Archer M (2005) Structure, agency and internal conversation. Cambridge University Press
Chronéer D, Ståhlbröst A, Habibipour A (eds) (2019) Urban living labs: towards an integrated understanding of their key components. Technol Innov 9(3):50–62
Ellen RF (1984) Ethnographic research. A guide to general conduct. Academic Press
Fava F (2017) In campo aperto. L'antropologo nei legami del mondo. Meltemi
Fava F (2021) Illusion of immediate knowledge or spiritual exercise? The dialogic exchange and Pierre Bourdieu's ethnography. In: Biscaldi A, Matera V (eds) Ethnography. A theoretically oriented practice. Palgrave Macmillan, pp 129–158. https://doi.org/10.1007/978-3-030-51720-5_6
Fricker M (2007) Epistemic injustice: power and the ethics of knowing. Oxford University Press
Geertz C (1973) The interpretation of cultures. Basic Books
Hallauer E (2015) Habiter en construisant, construire en habitant: La 'permanence architecturale', outil de développement urbain? Métropoles, Politiques Urbaines Alternatives 17:1–19
Honneth A (2005) Invisibilité: Sur l'épistémologie de la reconnaissance. Lavoisier 129–130:39–57
Hossain M, Leminen S, Westerlund M (eds) (2019) A systematic review of living lab literature. J Clean Prod 213(10):976–988
Marvin S, Bulkeley H, Mai L, McCormick K, Palgan YV (eds) (2018) Urban living labs: experimenting with city futures. Routledge
Nesti G (2018) Co-production for innovation: the urban living lab experience. Policy Soc 37(3):1–16. https://doi.org/10.1080/14494035.2017.1374692
Ricœur P (2004) Parcours de la reconnaissance. Trois études. Seuil
Sawyer KR (2005) Social emergence: societies as complex systems. Cambridge University Press. https://doi.org/10.1017/CBO9780511734892
Stocking G (1983) Observers observed. Essays on ethnographic fieldwork. University of Chicago Press
Thompson S (2006) The political theory of recognition. A critical introduction. Polity
van der Brien B, Owen D (2012) Recognition and power. Axel Honneth and the tradition of critical social theory. Cambridge University Press
Young IM (1997) Justice and the politics of difference. Princeton University Press

Chapter 7
1,460 Days of Love and Hate: An Ethnographic Account of a Layered Job

Paolo Grassi

Abstract Building on four years of ethnographic fieldwork carried out in the office of the Mapping San Siro action-research group (Department of Architecture and Urban Studies, Polytechnic University of Milan) in one of the main social housing neighbourhoods of Milan, in this contribution I will investigate the role and meaning of the Urban Living Labs (ULL) from an 'internal' perspective. An ongoing process of building relationships and caring for a space has allowed me to develop a reflection on multifaceted dimensions of daily life in the neighbourhood. Moreover, through anthropological literature, I will critically analyse the frustrations often experienced by researchers involved in fieldwork and planning. These frustrations highlight issues that go beyond the neighbourhood, showing the territorial dimension of the space. I will then highlight some ethical implications as clues that offer a more grounded understanding of daily life, rather than solving those implications with ready-made answers.

Keywords Urban Living Lab · Fieldwork · Scalarity · Frustration

P. Grassi (✉)
University of Milano-Bicocca, Milan, Italy
e-mail: paolo.grassi@unimib.it

N. Aernouts et al. (eds.), *Urban Living Lab for Local Regeneration*, The Urban Book Series, https://doi.org/10.1007/978-3-031-19748-2_7

7.1 Introduction: Just a Usual Day of Work

15 February 2021.

It's 8:30 a.m. I park my car close to Off Campus, the office of the Polytechnic University of Milan I have been managing since April 2019, together with the action-research group Mapping San Siro, in one of the biggest social housing neighbourhoods of Milan.[1] In 2019, we installed a wooden table and benches donated by a Swedish company at the entrance, in collaboration with the municipality and other local actors. I wonder whether their Nordic design goes well with the rationalist architecture of the concrete blocks of the neighbourhood. Over the last few months, the benches have become a socialization place for groups of residents who use this public area to meet and play cards. But damn! Rubbish is everywhere.

I raise the grey shutters of Off Campus, take a bag and gloves, and go out to start my collection: beer bottles, cans, torn cards, and plastic bags with food residue. A resident who passes by complains about the noise and disorder that we have promoted through our intervention. Another asks me if I am paid to collect waste.

I return to the office, tidy up the desk, and sit down. I correct a couple of student papers and wait for Irene, an anthropologist who is going to collaborate with us for a few months. In the meantime, Ida, project manager for the Polisocial programme of the Polytechnic,[2] also joins me. She sits at the head of the table, raises her laptop screen, and immerses herself in her work. We all wear masks and observe social distancing due to the ongoing pandemic. Irene arrives on time. We sit looking at my computer monitor and call Stefano, the third anthropologist of the group, via Skype. Irene wants to talk to us about some research that has just begun on some foreign communities in the neighbourhood.

At 10:30 a.m., our weekly team meeting begins, online again. Our discussion focuses on the management of a communication campaign linked to Covid-19 screening offered to the residents of San Siro by a Milanese hospital, the University of Milan, and ALER (the Lombardy Residential Building Company), which manages most of the apartments of the neighbourhood. The communication campaign, promoted by a network of local subjects, guaranteed a good number of tests among the residents. From specifically operational aspects concerning the distribution of leaflets and the organization of appointments, our discussion moves to a more reflexive level. We ask ourselves about the meaning of our role in this public

[1] San Siro comprises 6,135 dwellings and a population of approximately 12,000 residents. It was built during the 1930 and 1940s to provide accommodation for working families employed in the local factories. It is now inhabited mainly by a young and foreign population (largely from Morocco and Egypt—see Cognetti and Padovani (2018) and Capanema Alvares in this volume). The Great Recession of the late 2000s, which erupted with the financial crisis of 2007–2008, worsened the living conditions of the residents, exacerbating the socio-economic polarization between San Siro and other parts of the city (see Comune di Milano and Assolombarda 2018).

[2] The programme aims to 'place the university in close contact with the dynamics of change in society, extending the university's mission to social issues and needs that arise from the territory, on both a local and a global level' (http://www.polisocial.polimi.it/it/home/).

health action, and we evaluate the difficulties encountered, the frictions existing among the institutions involved. Meanwhile, my phone keeps ringing and ringing. This morning, I manage the switchboard that has been activated to give information and help people with reservations for the tests.

After lunch, we go out into the street. I walk around San Siro carrying leaflets. I reach a courtyard to say hello to a gatekeeper whom I know very well. He is sitting on a step, reading an advertising brochure in the shade of a tree. A young boy next to him performs the last technical tests with a bicycle on which a large bag has been mounted to deliver food. The gatekeeper says it is his first day on the job. Meanwhile, Raffaella[3] is waiting on the pavement for her drug dealer. On the phone, she shouts, asking him if he has tinfoil with him. She moves nervously, leaning on the parked cars. After a few minutes, he joins her with a friend. They enter the courtyard.

Back in the office, I call the coordinator of an educational project with whom we collaborate. We agree to programme some workshops which we will conduct in March at a local school.

At 5:30 p.m., I meet one of my students on Zoom. Meanwhile, Matteo, a photographer and a close friend of mine, joins me for a walk around the neighbourhood and to take some pictures. At 6:30 p.m., I get in the car and turn the key, but damn again! The battery is flat and the engine does not start. I make a few calls, but then I remember that there is a garage a few metres from our office. I ask the owner to help me out. The man comes with a briefcase and two electric cables. While accompanying me to the car, he explains that he has lived and worked in San Siro for thirty years. His words are full of resentment and anger. He blames the problems of the neighbourhood on the visible effects he perceives every day: for him, foreigners are the cause of the progressive deterioration of San Siro; that's it. 'Do you know that there was a shooting here on New Year's Eve?', he asks, pointing to a parking area in front of our office. 'The person who was shot in the head was lying here. What do you think can be done?', he asks me.

This lengthy field note is a good description of a typical working day in the Urban Living Lab Off Campus, located in the centre of San Siro. It represents a single day among 1,460 days[4] accumulated in four years as a postdoctoral fellow, carrying out ethnographic research on violence and urban space and supporting the activities of the action-research group Mapping San Siro.[5] I believe this field note strongly brings out two dimensions of my work, which I would like to reflect on in the following pages. The first has to do with the different levels on which my colleagues and I find ourselves operating. From the care for a physical space to the management of institutional relations, from participant observation to the coordination of a communication campaign, from applied activities to theoretical reflections, working in an Urban Living Lab has basically meant developing what I call a 'layered professionality'. In a circular process, this multiscalar profession has been poured into and reflected in my research activity.

---

[3] For confidentiality, apart from those of my colleagues, all names I mention are fictional.

[4] Up to January 2022.

[5] www.mappingsansiro.polimi.it.

The second dimension relates to a more methodological and ethical sphere not really problematized by anthropological literature, which questions my positioning as a researcher in an action-research group. While being significant and prominent, fieldwork often hides strong frustrations. What does dealing with these frustrations mean? How can they influence my work and the possibilities of change within the action-research process in which I am involved? The next pages therefore also represent a short story of 'love and hate', satisfaction and anger, successes and defeats; of contradictions inherent in any research process that aims to approach territories and their intrinsic ambiguity.

7.2 A Multilayered Job

Since its foundation in 2013, Mapping San Siro has progressively developed both research practice and theoretical reflection based on fieldwork (Cognetti and Ranzini 2016). Born as a workshop aimed at a group of urban planning students, the group was structured by accentuating its interdisciplinary nature, thanks to the permanent involvement of cultural anthropologists. Along this path, Mapping San Siro crossed the definition of the Urban Living Lab (Steen and van Bueren 2017), not only recognizing itself in it but also attempting further elaboration. In fact, the group currently defines itself as a *socially* oriented Urban Living Lab (Franz 2015), that is, a process-oriented space focused on social innovation, user empowerment, and codesign (Maranghi and Cognetti 2020). Other experiences implemented in the Milanese context starting from the early 2000s in some ways influenced the actions and reactions of Mapping San Siro and their recognition and distancing. For example, I refer to the Neighbourhood Laboratories[6] promoted by the municipality as part of an important redevelopment project (Neighbourhood Contract II) of five social housing neighbourhoods of Milan (see Cella 2006),[7] or to the Community Hubs promoted by the Cariplo Foundation since 2017. While the Neighbourhood Laboratories were intended by the municipality as 'front offices to inform and communicate with residents and to develop activities related to participatory planning' (Maranghi and Cognetti 2020, p. 99), the Community Hubs were especially oriented to urban regeneration (Calvaresi and Pederiva 2016). The research dimension, inherently part of Mapping San Siro, was thus lacking in these experiences.

Over the years, Mapping San Siro's approach has increasingly been defined around four elements that put in place—in a circular way—multilevel actions, all of which can be found in my field note. Taken together, the four elements concretize the idea of a socially oriented Urban Living Lab, crossing it with the action-research paradigm (Castelnuovo and Cognetti 2013). First of all, 'situating' can be defined as staying

[6] Laboratori di Quartiere in Italian.

[7] The programme was promoted in 2005 by the Ministry of Infrastructure and Transport, the Lombardy Region, and the Municipality of Milan.

in a specific territory, taking care of it, and building relationships from daily interactions and practices. Opening our office in San Siro[8] at least three days a week means reactivating an abandoned commercial space and stating the importance of social housing neighbourhoods through our work. Collecting rubbish and sweeping means assuming responsibility for a specific urban place, deconstructing the territorial stigma (Wacquant 2008), and claiming spatial justice against its marginalization (Philippopoulos-Mihalopoulos 2014). The second element of this approach, which also emerges in my field note, is 'networking'. Since 2014, the group has coordinated a network of local actors called *Sansheroes*, aimed at fostering a complex and deep-rooted knowledge of San Siro, useful for building shared visions of transformation (Maranghi 2019). Networking also means working to carry out collective actions and building relationships at a supralocal level, as I described using the communication campaign linked to the Covid-19 screening. The third element is 'inquiring', that is, conducting research through an interdisciplinary methodology and fieldwork.[9] My note immediately highlights some typically urban research themes, such as job insecurity in the case of the boy intent on preparing his bike to deliver food to homes; marginalization and drug dealing in the case of Raffaella; and intergenerational and interethnic relationships in the case of the mechanic I met before going home. Furthermore, I could easily add other themes, less directly linked to observation: urban and health policies, participation, public space, and so on. In Mapping San Siro's view, inquiring is strictly associated with teaching, in a circular way: 'The learning process is a hybrid process where research and teaching practice are connected, and the focus is on how the co-creation of knowledge can generate action and impact. A virtuous circle between practice—experiential and situated—and theoretical elaboration takes shape', wrote my colleague Ida Castelnuovo and Prof. Francesca Cognetti, the supervisor of the group, some years ago (Castelnuovo and Cognetti 2013, p. 41). Teaching constitutes an equally important part of our work (previously mentioned in relation to correcting papers in the morning and the Zoom call at 5:30 p.m.), which allows us to share our practices outside the group and reflect on them recursively and reflexively. Finally, the fourth element is 'acting', that is, the idea of engaging small 'levers of change' through our work, a dimension already inherent in the first three elements mentioned but which also acquires its own peculiarity, as testified by my phone call to the coordinator of the educational project with whom we collaborate.

The scheme presented, as exemplified through my field note, highlights the different levels on which we are called to operate as researchers engaged in a socially oriented Urban Living Lab. Working close to a territory, with different networks, teaching, doing research, and planning means implementing disparate actions that, from the everyday life of the neighbourhood, reach levels of broader

[8] The space was given in usufruct by ALER, the public company that manages San Siro.

[9] Mapping San Siro research was at first oriented on three main themes: home and dwelling; courtyards, public, and common spaces; and vacant and underused spaces. Other themes were added along the research directions of the individual social scientists involved. For example, I have personally developed a line of investigation into the relationship between violence and urban space, also crossing the three main themes of the group.

abstraction. The degree of organizational complexity and the skills required (communication, problem-solving, interpersonal skills, time management, and so on) define a hybrid professional figure that combines various fields of knowledge and practices such as managing an office, cultivating relationships with the residents of San Siro, coordinating institutional tables, and writing academic papers.[10]

Referring to this last point and further deepening the theoretical dimension of my reflection, it is worth considering how this 'layered professionality' is poured into and reflected in my ethnographic research. In fact, my entire investigation resulted in an ongoing process of building relationships that allowed me to open my own interpretations of multifaceted dimensions, inherent in the daily life of the neighbourhood as well as the policies and structural dynamics that govern it (Grassi 2018, 2019, 2020).

After all, cities have always questioned the traditional assumptions of anthropological work, as stated, for example, by Rivke Jaffe and Anouk de Koning:

> Urban landscape clearly extends beyond people's specific experiences and horizons, and is shaped by economic and political processes at the scale of the nation and the region. These features complicate the archetypical forms of anthropological fieldwork, based on long-term immersion in the socio-cultural life of a local community (Jaffe and de Koning 2016, p. 14).

The analysis of cities unequivocally calls for a multiscalar methodology, moving from a microsocial to a macrosocial approach and back. Urban space not only refers to the 'built environments' (Lefebvre 1974; Low 1990; Massey 2005), but is inscribed in their very fabric, their design, and their geographical distribution. A 'sense-experienced urban space' goes together with 'a non-ethnographic urban space', the urban space that we cannot see here and now (Fava and Grassi 2020). If this hypothesis is true for urban anthropology in general, it assumes even more value for anthropology practiced within an Urban Living Lab. Mapping San Siro's activity essentially demonstrates this: its progression establishes the requirement for working close to the territories, but it, in its attempt to be interpretative and applicative, also needs to go beyond them.

7.3 Dealing with Frustrations

In at least two points—when I cleaned up the space in front of our office and when I talked to the mechanic—the field note highlights a dimension that has been little considered by anthropological literature and by urban studies more generally, concerning the frustrations that fieldwork often entails. I certainly do not want to minimize the satisfaction and results achieved in these four years, but I want to reflexively show a grey zone often hidden by the rhetoric of committed research. The presence of waste in San Siro, for example, is a problem felt by most of my

[10] The hybridity of our professionality questions the theme of interdisciplinarity and the relationship between different researchers who, on the one hand, maintain their disciplinary specificity and, on the other, continually contaminate their gaze through collaboration (Cognetti and Fava 2019).

interlocutors. We need to refer to many causes in order to understand it: the low investment in the neighbourhood by some residents, due to the precarious housing conditions; insufficient institutional attention on the part of the municipality responsible for waste disposal; or the dumping of bulky items and debris by people outside the neighbourhood. However, beyond these explanations, cleaning the public space in front of our office every day is a hard job that puts into question our approach and goals. Why insist on taking care of such a small space in the neighbourhood? Are we perhaps imposing an idea of public space and 'urban decorum' (De Giorgi 2005) that does not correspond to that of the majority of the residents? Similarly, the issue of problematic interethnic relations in the neighbourhood is a 'wicked problem that cannot be approached from a single perspective' (Rittel and Webber 1973). But the complexity of the analysis collides with the words of the exasperated mechanic, whose only answer is racial hatred. How should we respond to this man's actual exasperation? Perhaps these questions would arise in any research, but they certainly emerge amidst the daily practice of a socially oriented Urban Living Lab, so close to planning and the dimension of change. These questions speak of anger, irritation, and impotence—'bad feelings' that we would usually not like to know in our idealized activity as experienced researchers.

More than other disciplines, contemporary anthropology has made reflexivity one of its fundamental characteristics. Clyde Kluckhohn (quoted in Remotti 1990), among others, spoke about anthropology as a mirror through which readers could reflect themselves. But what can we do when this mirror gives us back the worst part of us? Many ethnographies report anecdotally on situations in which anthropologists have experienced difficulties, but without making these an object of actual reflection. The introduction to *The Nuer*, published by Edward Evan Evans-Pritchard in 1940, is emblematic in this respect. Perhaps for the first time in the history of the discipline, the relationship with the interlocutors is publicly described in its fullness, also referring to the problems encountered in accessing the field: 'I arrived in Nuerland early in 1930. Stormy weather prevented my luggage from joining me at Marseilles, and owing to errors, for which I was not responsible, my food stores were not forwarded from Malakal and my Zande servants were not instructed to meet me' (Evans-Pritchard 1940, p. 9). And, further on 'Nuer are expert at sabotaging an inquiry and until one has resided with them for some weeks they steadfastly stultify all efforts to elicit the simplest facts and to elucidate the most innocent practices' (Evans-Pritchard 1940, p. 12). Since then, anthropology has certainly made great strides in including its interlocutors within the research process, questioning the power this exerts, and trying to smooth out its distinctive asymmetry (see Clifford and Marcus 1986). Throughout this process, the field has been thematized and problematized, also taking into account the frictions, contradictions, difficulties, and dangers it can activate. For example, some spoke of the dangers of fieldwork (Nordstrom and Robben 1996), some of boredom (Taussig 2004), and some—even those not strictly in the field of anthropology—have directly touched on the theme of frustration, although often just looking for strategies to fight it (see Goldsmith 2003; Parker 2001).

However, from the point of view of the researcher, what does dealing with frustrations mean and entail? First of all, the etymology of the word can come to our aid.

Frustration, from the Latin *frustratio*, means 'delusion', deriving from the verb 'to frustrate'. Frustration refers to those who believe that their actions have been in vain. In this sense, it is easy to see how the frustration experienced in the field is a fundamental alarm bell leading one to ask oneself about one's own advancement. However, what I would like to suggest is the need to accept frustration, to *remain* within it, to make it a reflective element on the basis of which to redefine the (action-)research process. Let me return to my field notes for the very last time. Let me shift the focus from my anger about the rubbish and my discomfort with the words of the mechanic to the moment of the group meeting in which instead we were able to recognize the frustration and make it a resource. I refer to the moment at which, analysing the problems encountered in the management of the screening, our discussion opened up to a new dimension. 'We ask ourselves the meaning of our role in this public health action, we evaluate the difficulties encountered, the frictions existing among the institutions involved'. Our actions collided with problems, making us think that our efforts were in vain. However, these problems were accepted by us, integrated into a horizon of meaning that led us to advance interpretations on themes such as the quality of institutional presence in the marginal neighbourhoods of Milan, the role of local knowledge in the activation of the community, the importance of territorial networks, and so on. Basically, expanding our gaze beyond the local context allowed us to look at more structural issues, thus relativizing the frustrations that emerged through our fieldwork.

7.4 Conclusions

Anthropologist Arjun Appadurai, together with media researcher Neta Alexander, recently published a book entitled *Failure* (Appadurai and Alexander 2020). Although the volume specifically describes how some forms of breakdown operate in financial and technological systems, a 'heterodox' reading of it can provide interesting insights into the somewhat analogous concept of frustration. Consider the first paragraphs of the text, where the authors write: 'We believe that the sense of failure is real and that it produces disappointment, regret, remorse, and other costly effects on individuals and groups. [...] [It] is a product of judgments that reflect various arrangements of power, competence, and equity in different places and times' (Appadurai and Alexander 2020, p. 1). It is difficult not to find similarities—even if not merely with reference to research practices—with what I have described so far.

In order to analyse this category, Appadurai and Alexander (2020) mention four 'schools of thought' that help them think about it in a creative way: science, queer studies, business, and infrastructure studies. While business and infrastructure studies have no direct connections with my analysis of frustration, the other two fields are worth mentioning—even very briefly. As is well known, modern science is based on the failure and the idea of refutability: 'An experiment is successful if it refutes a false hypothesis and forces scientists to come up with a new, and often better, explanation for the same phenomenon', state the authors (Appadurai and Alexander 2020, p. 4). Failure, like frustration, can indicate the direction not to follow, the attempt to modify, and the path to rework. This apparently simple observation is too often forgotten, especially in the 'regenerative' and 'empowering' world of planning, so conditioned by the rhetoric of success. Going beyond this rhetoric, Appadurai and Alexander (2020) mention the field of queer studies. Authors such as Ahmed (2010) and Berlant (2011), for example, invite a more nuanced comprehension of some dominant beliefs of our neoliberal era, primarily the so-called 'toxic positivity', that is, the idea the people should maintain a positive attitude in any situation.

The description of what, within a socially oriented Urban Living Lab, I have defined as a layered professionality (with respect to the skills requested and the type of actions implemented) focused on the need to deal with 'bad feelings', such as frustrations, as one of its inherent characteristics, beyond a predictable rhetoric of success. Loving and hating fieldwork (or, better to say, the satisfaction and delusions inherent in it) are part of the game. However, recognizing and accepting frustrations, or *remaining* within them, also by formulating alternative explanations, could take on an ethical value, a specific political positioning—political because critical of some dominant forms of thought—within the planning and more general academic world, I would suggest.

References

Ahmed S (2010) The promise of happiness. Duke University Press

Appadurai A, Alexander N (2020) Failure. Polity Press

Berlant L (2011) Cruel optimism. Duke University Press.

Calvaresi C, Pederiva I (2016) Community hub: Rigenerazione urbana e innovazione sociale. In: Bidussa D, Polizzi E (eds) Agenda Milano. Ricerche e pratiche per una città inclusive. Fondazione Feltrinelli

Castelnuovo I, Cognetti F (eds) (2013) Fare didattica sul campo, un anno di sperimentazione. I Quaderni di Polisocial n.1. Fondazione Politecnico

Cella M (ed) (2006) Un contratto per la città. I contratti di quartiere II a Milano. Edicom Edizioni Gorizia

Clifford J, Marcus GE (eds) (1986) Writing culture: the poetics and politics of ethnography. University of California Press

Cognetti F, Fava F (2019) La città interdisciplinare. Per itinerari non tracciati tra saperi urbani. Tracce Urbane 6:6–21

Cognetti F, Padovani L (2018) Perché (ancora) i quartieri pubblici. Un laboratorio di politiche per la casa. Franco Angeli

Cognetti F, Ranzini A (2016) Mapping San Siro. Strumenti di ricerca-azione nel/con il quartiere San Siro a Milano. I Quaderni di Polisocial n. 4. Fondazione Politecnico

Comune di Milano and Assolombarda (2018) Osservatorio Milano 2018. Comune di Milano e Assolombarda

De Giorgi A (2005) Dalla tolleranza zero al decoro. Dinamopress, 6 November 2005. https://www.dinamopress.it/news/dalla-tolleranza-zero-al-decoro/

Evans-Pritchard EE (1940) The Nuer: a description of the modes of livelihood and political institutions of a Nilotic people. Clarendon Press

Fava F, Grassi P (2020) Violence and space. A comparative ethnography of two Italian 'badlands'. ANUAC 9(1):183–210
Franz Y (2015) Designing social living labs in urban research. Info 17(4):53–66
Goldsmith A (2003) Fear, fumbling and frustration: Reflections on doing criminological fieldwork in Colombia. Crim Just 3(1):103–125
Grassi P (2018) L'angosciosa resistenza: Decostruire la categoria dell'«abbandono istituzionale» nel quartiere di edilizia popolare di San Siro (Milano). Archivio Antropologico Mediterraneo 20(2):1–18
Grassi P (2019) When interlocutors die: time and space of mobility through the biography of a homeless man. Ethnography. Advance online publication. https://doi.org/10.1177/1466138119891453
Grassi P (2020) 'Puliamo San Siro': Lottare contro lo stigma territoriale in un quartiere di edilizia popolare di Milano. Archivio Antropologico Mediterraneo 22(2):1–15
Jaffe R, de Koning A (2016) Introducing urban anthropology. Routledge
Lefebvre H (1974) La production de l'espace. Anthropos
Low SM (1990) The built environment and the spatial form. Annu Rev Anthropol 19:453–505
Maranghi E (2019) Il ruolo della conoscenza locale nell'attivazione di comunità di pratiche e di progetto. Il caso della rete Sansheroes (Milano). Tracce Urbane 3(6):177–195
Maranghi E, Cognetti F (2020) Milan. In: Aernouts N, Maranghi E, Ryckewaert M (eds) Towards a definition of socially-oriented living labs. SoHoLab, pp 93–116
Massey D (2005) For space. Sage
Nordstrom C, Robben ACGM (eds) (1996) Fieldwork under fire: contemporary studies of violence and survival. University of California Press.
Parker KC (2001) Enrichment and frustration in fieldwork. Geogr Rev 91(1–2):168–174
Philippopoulos-Mihalopoulos A (2014) The movement of spatial justice. Mondi Migranti 8:7–19
Remotti F (1990) Noi, primitivi. Lo specchio dell'antropologia. Bollati
Rittel HW, Webber MM (1973) Dilemmas in a general theory of planning. Policy Sci 4(2):155–169
Steen KYG, van Bueren EM (2017) Urban living labs: a living lab way of working. AMS Research Report 2016–2017. AMS Institute
Taussig M (2004) My cocaine museum. University of Chicago Press
Wacquant L (2008) Urban outcasts: a comparative sociology of advanced marginality. Polity Press

Chapter 8
The Inside and Outside of High-Rise Social Housing: The Broken Institution

Simon Allemeersch

You may complain about how our towers look from the outside, but we have the view.

Abstract This article is based on an ethnographic report of a long-term artistic workplace in the inside world of a social high-rise ensemble. The communal 'atelier' functioned as a repair workshop (Allemeersch et al. 2014) concerned with re-negotiating the relation and knowledge between the inside and outside world. Through a synchronic ethnographic report on the inside world of the housing ensemble, this article aims to characterize the lived citizenship (Warming and Fahnøe 2017) of residents, formal and informal, based upon the opposition between formal and informal order (Goffman 1966; Scott 1998), the notion of 'underlife' (Goffman 1963/1996) and hidden and public transcripts (Scott 1990). As the deserted stronghold of a previously 'pillarized' welfare state, this article pictures an inside world that is unbalanced between formal and informal order, and lacking the latter (Scott 1998). This results in a social closure between the inside and outside world, and the loss of self of residents. Essentially, residents are caught in the double bind between isolation and social closure (Wacquant 2008) on the one hand, and the loss of façade (Goffman 1959/2019) on the other. Without a qualitative understanding of the inside world of high-rise social housing the outside world institutions act without any knowledge of 'the community that many residents were able to create in such adverse conditions' (Goetz 2011, p. 270), and the difficult relation these residents have developed towards their own environment and housing, state intervention and the public services (Wacquant 2008).

Keywords High-rise social housing · Marginalization · Formal and informal order · Incapable tenant · Urban ethnography

S. Allemeersch (✉)
Department of Social Work and Social Pedagogy, Ghent University, Ghent, Belgium
e-mail: simon.allemeersch@ugent.be

N. Aernouts et al. (eds.), *Urban Living Lab for Local Regeneration*,
The Urban Book Series, https://doi.org/10.1007/978-3-031-19748-2_8

8.1 Introduction

When I started working in the social high-rise ensemble of the Rabot neighborhood in Ghent (Flanders, Belgium), a janitor told me that the only communal space provided when the three towers were finished halfway the 1970s was a mortuary in the basement of the second tower. The first time I heard this, I thought it was a grim joke. Next to apartment 019, which was a meeting place run by social workers and residents in the third tower, my workspace in the first tower was the only communal space. The idea of organizing a workspace started in 2009 when it was officially announced that these towers would be torn down. One year later, in November 2010, I was able to occupy it. Deliberately having no preconceived project, I was curious what would happen if I decided to do all my work as a theatre maker from that location. In 2012, I moved out of the first tower with the last two residents, and moved the workspace to the second tower, which was then still fully occupied. Officially, the atelier closed in February 2013, but I continued to work with this network of residents up to 2014, when a theatre performance and a book about these towers premiered at the Vooruit Arts Centre (called Rabot 4–358; Allemeersch et al. 2014). I kept in touch with residents and ex-residents—and organized working periods in the second tower and the third tower—up to 2018. The research was gradually structured into (ongoing) doctoral research. In June 2020, the last residents left the third and last tower, which was finally torn down in spring 2021.

Throughout the years, the workspace had been an atelier and working place for different artists, a meeting point, a kitchen, a welcoming shelter for informal residents, a workspace for researchers, a meeting point of an arts festival, a coffee table after a funeral, a small cinema, an exhibition, etc. At the back of the apartment, I had put up a wall-to-wall picture of a theatre stage, an *infini* of a forest. At night, when you stood outside in front of the first tower, you could see a forest through the window (Fig. 8.1).

8.2 Urban Narratives and Symbolic Violence

Revealing the existence of an insider's perspective on social housing and an outside world surrounding these communities may seem self-explanatory. Yet, my goal here is to characterize the opposition between insiders' experiences and the images the outside world tends to project on the communities within the towers. Inside-world problems of social housing residents often differ from the outside world's discourse around social housing: Sahlin (1996, p. 168) argues that the public definition of social housing problems is partly independent of what residents may experience or think. These same residents have been cast as 'the incapable tenant' (Sahlin 1996), 'labelled as deviant, noisy, dirty, dependent, ungrateful and above all, incapable of living together with other people' (De Decker and Pannecoucke 2004, p. 294). Yet, without a qualitative understanding of the inside world of high-rise social housing, the

Fig. 8.1 The infini at Rabot 1

outside world's institutions act without any knowledge of 'the community that many residents were able to create in such adverse conditions' (Goetz 2011, p. 270), and we risk missing a deeper understanding of the difficult relationships these residents have developed with their own environment, with state intervention, and with public services (Wacquant 2008).

Therefore, Vervloesem (2019) points at the agency of these urban narratives and their epistemological relevance to urban planning: 'The (re)writing of stories about the city is, in an epistemological view, a necessary step to look critically at the role of urban planning', which brings us 'to another, more diverse and layered scope of action for the urban planner' (own translation 2019, p. 42). Vervloesem thereby cites Leonie Sandercock: 'The way we narrate the city becomes constitutive of urban reality, affecting the choices we make, the ways we then might act' (2019, p. 42). This epistemological awareness in studying the complexity of urbanism contextualizes knowledge as a 'lived experience' (Warming and Fahnøe 2017), 'as a part of human experience and as growing out of human practice' (Harvey 1973, p. 296). This is not necessarily the same as a unifying and simplistic counter-discourse against dominant social groups, because 'different power relations may also be at play within marginalized groups, and "the government" may be a multitude of contradicting voices' (Vervloesem 2019: 44—own translation).

Both Dehaene (2008, 2011), Vervloesem (2019) have drawn attention to the work of urban sociologists Jean Remy and Lilliane Voyé and their understanding—in *Ville, ordre et violence* (1981)—of the interstitial space in the dialectic relation between *first order* and *second order* spaces. Remy and Voyé (1981) state how a heightened perception of symbolic violence can arise when this interstitial space cannot function in between the functional efficiency of the first order and the less coded, uncertain, fluid spaces of the second order. This interstice is a 'space of communication that serves to construct collective identities [and] thus introduces a better understanding of local and regional dynamics' (Remy 1993, p. 272, as cited in Vervloesem 2019, p. 133). It is part of a larger 'ecology of weak integration', 'a determining factor for the capacity of an urban environment to accommodate different social groups and their divergent ambitions' (Dehaene 2011, pp. 1–2), in which Remy finds a transition 'from urban configurations which consist out of [sic] spaces that could have changing meanings over time, to an increasingly specific and specialized world in which singular places have singular functions' (Dehaene 2011, p. 12).

Remy and Voyé argue that urban configurations tend to allocate specific groups to specific places (Remy and Voyé 1981, p. 103), which results in reduced spatial indeterminacy, and therefore, 'reduced "room to play" for those groups that exist off the radar, those groups that do not fit the categories explicitly catered for'—in particular the disenfranchised groups. This leads 'to the suppression of the opposition of front and back stage' (Dehaene 2011, p. 12). Hence the symbolic violence in the title *Ville, order and violence*, which is imposed on these groups.

8.3 Erving Goffman and Informal Order

Perhaps Erving Goffman is mostly known for his dramaturgical theory (Goffman 1959/2019). Historically, the *theatrum mundi* analogy is often distrusted (Rancière 2010; Sennett 1978). But Goffman used this analogy to explain face-to-face social interaction between individuals and has often been misquoted[1]: his dramaturgical theory does not state that individuals are merely actors, or that the realities they experience are part of an illusional make-believe world, as Goffman made clear himself (Goffman 1974/2004, p. 1). Humans live an often all too real reality, albeit we frame this experience differently. The different ways in which individuals realize their social *self* within social interaction through impression management can best be understood using the terminology of actors working on a theatre stage (Goffman 1959/2019, p. 270).

Therefore, the work of Erving Goffman offers a valuable framework for an ethnographic perspective on face-to-face interaction, micro-sociology, and the effect of institutional frames on the individual self. Wacquant, with a motto to his ethnography *Body and Soul* (Wacquant 2004, p. 2), pays his respects to the ethnographer Goffman: 'Any group of persons [...] develop[s] a life of their own that becomes meaningful, reasonable and normal once you get close to it', (Goffman 1963/1996). Goffman's analysis of stigma (1963/2018) remains one of the cornerstones of Wacquant's oeuvre (Wacquant 2019) and is still relevant to critical thinking on social work (Garneau and Namian 2017) or the analysis of the coping strategies of marginalized individuals (Rondelez et al. 2018).

[1] Too often, Goffman's dramaturgical theory is linked with the partially quoted lines from Shakespeare's 'As you like it'. The quote has an entirely different meaning, and Goffman himself rebukes the reference in the introduction of *Frame Analysis* (Goffman 1974/2004).

Goffman's (Goffman 1959/2019, p. 66) *façade regions* can be understood as any space that is bounded to some degree by barriers to perception and which offers the necessary boundaries to perform the roles of social identities. Goffman defines 'social institutions' as spatially bordered and shielded from the outsider's gaze (Goffman 1959/2019). This leaves open ample analysis opportunities, outside of the classic examples of asylums or prisons and their inmates: 'A social establishment is any place surrounded by fixed barriers to perception in which a particular kind of activity regularly takes place', and therefore, 'any social establishment may be studied profitably from the point of view of impression management' (Goffman 1959/2019, p. 252).

Throughout his ethnographic works, Goffman's main focus has always been on the realization of the social self through social interaction, be it in everyday life (Goffman 1959/2019), in the extraordinary circumstances of stigmatized or mentally troubled individuals, or in the lives of the inmates of total institutions (Goffman 1963/1996, 1963/2018, 1963/1966). Goffman defines *stigma* as a disqualifying social attribute, regardless of its intrinsic value, that weighs upon the social relation and prevents the stigmatized person from realizing the intended social identity. This leaves the stigmatized person with a *spoiled* identity (Goffman 1963/2018). The individual that is subjected to a *total institution* risks losing the notion of *self* and *mortifies*, having nothing left but to accept the ruling vision that the social institution imposed upon his or her social identity (Goffman 1963/1996, p. 134). A critique of Goffman would picture humans as being overdetermined by the social relation and a totalitarian power (for this discussion, see Rondelez et al. 2018), but one can argue that throughout his work, Goffman always leaves room for the virtuosity and gameplay of individuals: his self-acknowledged attachment to players, cheaters, and the like (Goffman 1974/2004) makes clear how much he values the relevance of a non-formal order and of individual agency. Goffman refers to the non-formal not as 'informal', because he reserves the notion of 'informal' for the looser customs within the formal order. (He is not easily fooled: *your boss asking you to go out for a drink* may seem informal, but remains part of the formal order.) The set of rules, customs, and communication that does lie outside the formal itself is the *underlife* of institutions (Goffman 1963/1996), in which the inmates find roles, contacts, and powers that are not permitted by the formal order in itself but through which these individuals may escape these powers.

I hereby rely on the opposition between formal and informal order as it was described by Scott (1998), and I understand this informal order as the *underlife* or the unofficial order as denoted by Goffman (1963/1996, p. 140). In a critique of high-modernist schemes and formally organized social action (which includes the modernist CIAM high-rise architecture), Scott (1998) points to the simplifying formal rules that can never generate a functioning community. Therefore, these formal schemes remain inadequate 'for creating a successful social order'. Scott continues: 'Formal order, to be more explicit, is always and to some considerable degree parasitic on informal processes, which the formal scheme does not recognize, without which it could not exist and which it alone cannot create or maintain' (Scott 1998, p. 310). We can also trace this relevance of the informal order for the urban environment in the work of Richard Sennett (2018)—already prepared in his earlier work (*The Uses of Disorder*, Sennett 1970)—and ultimately to Jane Jacobs: 'Formal public organizations in cities require an informal public life underlying them, mediating between them and the privacy of the people of the city' (Jacobs 1961, p. 57). The informal order renders services that 'cannot be formalized', such as mutual trust among residents of public housing projects (Jacobs 1961, p. 60).

Finally, I remark how Scott, in his *Weapons of the Weak*, makes a Goffman-like distinction between *onstage acts* and *off-stage* activities (Scott 1987, p. 41), which results in his later analysis of 'public' transcripts (Scott 1990) as the hegemonic 'self-portrait' of the dominant elite (p. 18). This is opposed to the 'hidden transcript': a whole of stories, gossip, conduct, and off-stage activities of the dominated groups, not just 'evasive but often inaccessible altogether' (p. 19).

8.4 Then What is This All About?

In this article, I will focus on the communities in the social high-rise ensemble of the Rabot towers in Ghent as seen through the daily activity of an on-site artistic workspace (Allemeersch et al. 2014). The active observation period ran from November 2010 to February 2013, and this article focuses on a more synchronic description and ethnographic report of the small worlds that resided in these towers during this period. This qualitative research is based on the field notes gathered during the observation period, a book published about this period (Allemeersch et al. 2014), and semi-structured interviews with participants and professionals working in social housing.

In qualifying the opposition between the formal and the informal order, the shifting symbolic borders between the inside and the outside of social institutions, and the hidden transcripts through which individuals handle unequal power relations, I will show how these oppositions reveal the specific situation that marginalized social high-rise residents find themselves in, marked by 'the spatial denigration [...] that affects the subjectivity and the social ties of their residents as well as the state policies that mould them' (Wacquant 2019, p. 40).

8.5 Findings

8.5.1 The Loss of Façade

The small communities in the social high-rise ensemble in Rabot are much more heterogeneous than is suspected from the outside. As we find in Wacquant (2008, p. 172): 'What appears from the outside to be a monolithic entity is experienced by its members as a finely differentiated congeries of "micro-locales" centered on buildings and even on different stairwells inside the same building'. Residents are brought together in a divided way while being treated as one whole by the outside world. The heterogeneity of their world is not known or not sufficiently recognized. Participants feel condemned to each other, even more than to the place itself, and maintain 'micro-hierarchies' (Wacquant 2008, p. 183) within this inside world: 'This building doesn't work, but our floor is doing just fine' (participant B3a). Or: 'In the third tower things are really bad'. They often blame the collective shame on a specific infamous category of fellow residents (Wacquant 2008, p. 183). In Rabot, the graffiti in one of the elevators read: 'Kill pedos Kill moslims [sic]'.

From the participants' view, their isolation appears to have a double, paradoxical meaning: they experience a loss of symbolic façade combined with social isolation. Both are aggravated by the physical building itself. This becomes especially clear in the lack of acoustic insulation, which not only hinders the daily life practically (the usual quarrels about loud neighbors) but also makes it impossible to establish a façade amongst neighbors. The walls in Rabot were infamously thin, and residents could hear their neighbors in all directions. As a consequence, 'all the neighbours knew they know too much' (Goffman 1959/2019, p. 129). This lack of acoustic insulation worked in one other specific way in Rabot: the intercoms in the entrance hall were broken, and they never disconnected. Private sounds, such as household arguments, private conversations, or children crying, could always be heard in the entrance hall.

The loss of façade also occurred in relation to the outside world, e.g. the feeling of shame residents felt in front of family members or outside-world professionals (e.g. medical personnel) who visited the building and with whom residents had to share a dirty elevator (B3a, B3b, B7). But a symbolic façade is about more than just the concealment of private life.[2] It is also about the possibility of a public appearance on the individual's own terms, as this is the pre-requisite for the construction of a social identity (Goffman 1959/2019). Quickly, the atelier was used by residents as a façade region for the necessary negotiation between inside and outside—between private isolation and a much needed more public appearance:

> D. regularly visits the atelier in the morning, casually bringing and opening the mail he received that morning. He then reads the mail out loud, which is often news from his probation officer, or is about the alimony settlement. Then he can give his opinion, e.g. cursing the justice department. When the atelier is closed, he goes to the social worker and 019 meeting place – for the same small act. The social worker says he is 'too afraid to open his mail on his own' (own translation).

This lack of symbolic façade was also experienced because the only contact residents had left was with professionals, who gathered there only because of social problems. These professionals held a lot of private information about the residents. Residents were isolated but lived amidst an abundance of welfare organizations,

[2] For a critique of dealing with privacy in planning and a more subtle understanding of this as a balance 'made up of small, sensitively managed details, practiced and accepted so casually that they are normally taken for granted', see Jacobs (1961, p. 59).

which led to a general *fatigue* regarding welfare and social work. Although the question 'Are you a social worker?' always made me feel uncertain in the beginning, it turned out that my exceptional position was in my advantage, because new roles became possible.

8.5.2 The Informal Order

Whereas social workers for the housing company could still be considered street-level bureaucrats (Allemeersch 2022a; Lipsky 1980), in a difficult split between their official task and the playing rules of the field (Sahlin 1996), the janitors were key figures in this inside world. Being employees and tenants at the same time, the border between their official task and their status as residents was blurred. Different janitors from different periods also described the heavy mental weight of the job in the building where they lived. This was combined with hardly any specific selection or training for the job: 'The training was one day, showing you around and showing where the expansion vessel was' (B7a).

These janitors held informal power, next to their official responsibilities. Some janitors regularly asked for additional 'payment' from residents for completing their official task. 'They wouldn't move if not…' (participant makes a gesture of money between index finger and thumb, B4). I knew I had become an insider when a janitor asked me for this informal payment, because I needed help after locking myself out of the apartment.

One janitor was convicted for cheating an older lady out of her money when he went shopping for her groceries with her credit card. He gave this lady the receipt, but she could not read. This janitor was convicted for a proven 20,000 euros (Luyten 2014) and lost his job but kept his apartment. Officially, he was no longer a janitor, but informally, he continued to exercise his role as key figure. In an interview with an art collective that worked in the third tower from 2017 onwards, this same janitor was still regarded as being a most important key figure, although the project occurred three years after his dismissal. In the reality of the inner world, his dismissal made no difference: he still exercised the same powers and occupied the same role in the inside world. And even more: he devoted himself all the more to organizing activities through city neighborhood subsidies.

Another participant recalled how the widow of a janitor 'inherited' the knowledge and memory of her late husband and informally held a similar position among the neighbors 'because she knew a lot' (B11). This inside knowledge was crucial, as social tenants dealt with a landlord who looked more closely to private life:

> One older couple always visits the workplace together. Both are widow and widower of separate marriages, and both have their own apartment – but since years they form a couple. Out of fear of losing the luxurious situation of the two nearby apartments in the same tower, they hide the relationship.

Because of this pressure on private information, retreating in the hidden transcript of an inside world may not be that surprising, 'for suspicion and fear of trouble often outweigh any need for neighbourly advice and help. For these families the sense of privacy has already been extensively violated' (Jacobs 1961, p. 67).

8.5.3 A Morbid Atmosphere

Participants often referred to what I can only describe as a morbid atmosphere. For several reasons this may not be that surprising: old age and weak health of some residents, several incidents of drug-related, sexual, or domestic violence, or the risks and consequences of addiction in itself. In evaluating the work of the atelier, one participant gave an uncomfortable compliment:

> What you did was great. The simple fact that you were there for years and never had a fight with someone. No black eyes, no window thrown in – in that place… You know that that wasn't evident at Rabot (B10).

Violence, and the fear of violence, often became part of the default situation. In addition, the towers were infamous for several suicides, since they can easily be entered without keys and the staircase to the roof is always open because of fire regulations. Most participants witnessed one of these suicides directly: 'He wore a raincoat and it was like a balloon. Before I could say "What's that?" he was already on the ground. Blink with your eyes and he was down' (B7b). This person's husband immediately had to call a doctor, because she 'saw one coming down' (B7b).

One participant declared to have been asked by the police to identify a body in order to find out whether the deceased person was a tenant (B3a). These suicides, together with sometimes violent incidents caused by the presence of informal residents (Allemeersch 2022b), reinforced the divide between the inside world and the outside world: for the outside world, these incidents were part of the infamous character of the place, while residents themselves perceived these incidents as problems that came from the outside, for which they themselves could not be held responsible.

Next to these tragic incidents, the decision to tear down the buildings confronted residents with their own life expectancy:

> That was really a fear of death and that has a special scent. […] Not pleasant. But that was really there. Like there's suddenly more grey in the colours. People who weren't already pondering their death were suddenly all collectively in that block reminded about their death. People of my age, who knew they had to leave there in seven years, then they're in their seventies. Moving house at that age, am I going to survive that? And there were many people living there who were older (B10).

Care for isolated older residents was often absent, and together with failing building maintenance and unwanted visitors from outside, this affected many residents:

> A resident who passes away in the solitude of his own apartment during a heatwave, and is only discovered several weeks later, causes dismay. As well as the fact that the apartment is left uncleaned for another few weeks, then is opened up by squatters, and finally the fouled mattress of the man is left for several days in hallway on that floor (A6).

Although many of these incidents had causes in the outside world, these visceral experiences divided individual residents, as they caused some of them to disconnect from their immediate surroundings and their neighbors (B10).

8.5.4 Assistance of Organizations from the Outside World

When residents needed assistance, I noted a striking distrust in outside-world organizations. Even as one resident (B10) experienced insecurity and violence, she was clearly distrustful of police interventions, expressing concern for vulnerable tenants:

> So, you couldn't call the mobile team [mobile assistance team for psychiatric patients], you had to be a nurse, or a doctor. As a citizen, you had to call the police. But that was when the police had beaten a psychiatric patient to death in the cell [refers to the case, well known in Flanders, of Jonathan Jacob in 2010]. This happens, and you just don't call the police for people like that. The police are not trained for that. And I also think it's wrong to have psychiatric patients in a building like this, they need rest. I know about a patient who got a beating… (B10).

Participants attributed the possible success of these organizations to the merit of individuals, without any positive judgement of the organization itself. Statements like 'The police are scum, but this neighborhood officer was a good one'; '(*Name of social organization) is worthless, but (*name of social worker) was great' are very common. Residents truly understood Scott's observation of how formal order relies on informal personal knowledge and contact. But this meant that these organizations left no lasting positive impression. When there was a reorganization of the team, when a specific individual professional disappeared, the organization as a whole was disqualified.

As stigma and social identity are located in the social micro-relation and face-to-face communication (Goffman 1963b) or are only fully expressed within the hidden transcript, official institutions often remained unaware of the stigma, which they tended to downplay, having only 'a paper relation' with the residents, as described by one professional (A7). The divide between the housing association and its tenants often resulted in disciplinary reactions and strategies:

> After a traumatizing death of a close relative (who was a tenant in the same building), one resident puts up a letter in the hallway – a salute as well as an attempt at gossip. The letter is removed, because it is against regulations to put up messages in the hallway (B10).

Apart from distrust and a disciplining attitude towards tenants, I found a remarkable difference in the framing of the same reality by tenants and housing company officials.

> When the demolition was announced in 2009, the housing company organized a meeting with tenants to announce the demolition. For well-intentioned reasons of accessibility, the housing company chooses the auditory of a newly built university across the street, unaware of the intimidating effect a new university has on residents of the neighbourhood, who are not that 'schooled'. When the residents enter the auditorium, they are met with police presence, because the housing company asked for assistance beforehand.

If we compare the report of the director of the housing company (who led the meeting) and the report of one engaged resident about this event, we notice an important difference. The director expressed disappointment at so few residents showing up, which remained a mystery for him (Allemeersch et al. 2014, p. 170), while the resident described the same event and said that the venue was 'packed', stressing the fact that 'a lot of people showed up' (B3a). The director compares the number of people who showed up with the official number of tenants living in the entire ensemble. The resident had been engaged in tenant meetings for years and volunteered in social work. Therefore, he was pleased with the turnout, because he compared this with the difficult work over the past years.

8.5.5 Lost in Translation

Residents had a similar distrust of media and the arts, to whom they referred by one category: 'People with a camera'. This could have referred to journalists, photography students, or filmmakers. All too often, students of different academic backgrounds showed up with intimidating questionnaires. Ironically, for the workspace, this was the start of the interdisciplinary aspect of the work, because residents started to send these various students to the workspace ('You go and ask there…'). There was a remarkable curiosity from the outside world about these towers. When we organized a photography exhibition, visitors openly said 'we're not really interested in the pictures, but we just wanted to see these towers from the inside'.

My work as an artist and with theatre in general was greeted with curiosity by the residents ('How do you earn you money? Is this really your job?'). But some suspected a different reason for my presence, which reveals a lot about the distrust of the housing association:

> I am met with hostile looks in the elevator. Apparently, the gossip is that I am filming and interviewing old people on behalf of the housing company. On the basis of that material, the housing company will decide whether those people will get a new address, or will have to go to a home. When I'm interviewing A. at her apartment, a friend comes to warn her during the interview about my 'real' intentions: I'm here to see if she needs to go to a retirement home or not.

It should be noted that at that time, I had already written a letter to the residents, which could be seen and read everywhere. Apparently, as a sign, it was not powerful enough to clarify my background. At several moments, the lack of a habit of reading or literal illiteracy became clear:

> He entered the atelier with a sweater on which was written in big letters: ITALIA. He asked me what country the sweater was from. I asked him: how do you manage if you can't read? And he told me that he has his own system.
>
> Seven lines are the days of the week (Fig. 8.2). A small mark above the second line is Tuesday. When this mark is below the line it is the afternoon. He draws a small camera and says: that's you. I must write it as '3' instead of '15'. His address book is a series of hieroglyphs that represent the history of families, the sex of family members or ways to reach them (train e.g.)—combined with the telephone number (see Fig. 8.3). When there is a coffin drawn, or a line through it—then they have died. He says: I am not stupid, just not 'schooled'.

For the outside world, this illiteracy resulted in residents not having a lot of information and, for example, basing their information on gossip and hearsay. But illiterate residents themselves were also easy victims of gossip or abuse, as was already mentioned in the example of the janitor. When a vulnerable resident asked his neighbors to read his mail, because he was unable to, the residents learned about his past conviction. This illiteracy again echoes differently in the outside world, where the problem is too often uniquely attributed to the migration background of social tenants. But stating that the problems in the social high-rise had to do with

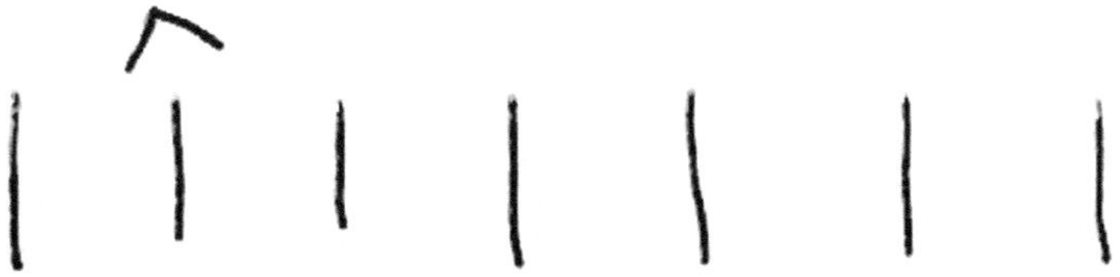

Fig. 8.2 Meeting on Tuesday morning

Fig. 8.3 Families and their telephone number

'the migrants who don't speak Dutch' is the ultimate way to be recognized as an unexperienced outsider in Rabot. For example, the Turkish community in the Rabot towers displayed strong social bonds, at least stronger than the network many of the white residents could rely on. These Turkish families had their own informal translators (often their children), whereas white residents combined their illiteracy with social isolation, which made their situation far more problematic.

Even within the framework of racist reasoning, participants acknowledged more urgent and pressing problems in the inner world of the building than cultural differences: foremost the psychological need and loneliness of some white tenants. As was said, 'Let's be honest: the biggest problems we had with our own kind' (in talking with B3a)—an opinion which denounces the typical racial slur which portrays tenants of color as the main source of all problems but acknowledges at the same time the difference between 'them' and 'our own kind'.

8.5.6 Pride, Knowledge, and Agency

A too deterministic view on these spaces would remain blind to the solidarity and care residents showed with and for each other and their surroundings. As impressive and emotionally disturbing as some of the incidents were, the daily life of residents displayed laughter and sometimes harsh humor while they kept the meeting place open or had morning gatherings in the hallway, waiting for the letter carrier. Residents took care, but they did this quickly and quietly. A loaf of bread was passed between neighbors in the hallway. People moved through their apartments without a sound, always thinking about the sick neighbor one floor down (B10). When there was a burglary in the workspace and even the copper cables of the speakers were stolen, a group of women came by. The next morning, I had all the cleaning material I needed and new speakers. Amid the bleakness of the alienating space, a lot of place-making was going on, alongside the gathering of social capital, as 'dis-identification did not mean withdrawal' (Blokland 2019, p. 54). During numerous interviews, residents talked about all the problems and always ended with concluding, 'I really like living here'. Pride was hardened by the conditions, rather than diminished. The attention in the newspaper for the demolition of these 'ugly towers'[3] only added insult to injury.

From the early start, the workspace linked up with the informal order and drew on the solidarity of residents and their network that held this knowledge ('If you want to know more about… go and see her'). But as we have seen, this commitment of residents was first of all part of the daily 'hidden transcript' of the inside world, concealed by its informal or illegal status. Or, second, the individual agency was shrouded in the singularity and isolation of the mentally troubled, and was easily misunderstood:

[3] The title of the article in the newspaper was 'Belated love for ugly towers', an article about the workspace and an exhibition that was organized. The article is reproduced in Allemeersch et al. (2014, pp. 222–223).

> At first glance, O.'s apartment is filled with garbage. Empty plastic bottles are everywhere. Only after a long time I learn that O. has heard a rumour about the government introducing a deposit on plastic bottles. He is focused on his plan: he's preparing to cash the deposit of these bottles.

As one social worker remarked: 'A homeowner can have the craziest hobby or obsession in his basement, but when our tenants dream about something, it's immediately a problem' (A11). 'The symptomatology of the "mentally ill" may sometimes have more to do with the structure of public order than with the nature of disordered minds', (Goffman 1963/1966, p. 242). Moreover, their agency is often problematized and seen as 'improperties' symptomatic of the mental illness (p. 232). To put it bluntly: only 'normal' people have the right to act out of the ordinary without repercussions.

8.6 Discussion

Ironically, an architectural typology that followed a high-modernist ideal of legibility (Scott 1998) produced a marginalized inside world which became largely illegible to the outside world. As such, residents were caught in a double bind between social isolation on the one hand and being exposed through the lack of symbolic façade on the other hand. The necessary formal order 'left the building', which left residents in the all too personal arbitrariness of an informal order. It made residents' agency difficult to notice, let alone support, for organizations or institutions were often isolated from the informal order of the inside world of these buildings and lacked the practical local knowledge, or *mētis* (Scott 1998). This broken formal order marked the social high-rise as a *broken institution*, with an effect similar to the one Goffman noticed within the total institution: the loss of self and mortification of its residents (Goffman 1963/1996).

Knowing this, outside-world initiatives should not only focus on the agency of marginalized residents, urging them again and again to participate. This is comparable to shouting 'don't be nervous' to someone who is nervous; it simply does not work. Rather, outside-world initiatives should try to acquire a 'critical lens' (D'Cruz and Jones 2004) on the circumstances under which the agency and knowledge of residents came about, and act upon this knowledge. The renegotiation between the outside and the inside world of these marginalized spaces has to include the question of to what extent, and how, the systemic outside world *participates in residents' life worlds* and whether outside-world projects are ultimately part of either 'the reproduction or the transformation of existing social relationships' (D'Cruz and Jones 2004, p. 9).

These effects on the residents of an unchecked informal order may be read as a warning against an all too naive understanding of the informal order as an answer to lacking institutional responsibilities. The formal and the informal order should be understood as a communicative balance. The interstitial space is not a counter-space, but a space of communication (Remy 1993) that has to be *marginal* in the best sense of the word: always relating to both inside and outside, to formal and informal (Vervloesem 2019).

Vervloesem (2019) points furthermore to the possible danger of essentializing the difference of the urban margin and marginalized residents, who are portrayed 'rather one-dimensionally as if they, as human beings, are essentially different from others and pathologically so' (Blokland 2019, p. 73). An outside world often understands the stigma as inherent to the stigmatized residents themselves. On this point, we reach the core of Goffman's interactionist view on stigma: the 'normal' and the 'stigmatized' are interchangeable roles (Goffman 1963/2018). The stigmatized person is by definition normal precisely because he or she strives to live up to the virtual and 'normal' social identity but fails and is failed in doing so, which results in a strained social relation, a spoiled identity, and the resulting stigma. This interactionist understanding of stigma is the main reason to always look at the inside *and* outside world of social housing when addressing the problematic status of social tenants. *Social high-rise residents are first of all normal people.*

Yet, finally, we are left with an ethical question, which applies to this very text and the research as a whole. To what degree may we disclose the hidden transcript and informal inside world—knowing that an individualist, essentialist reading often prevails in the outside world? *Probably not all of the outside-world public has the same view on stigma as Goffman.* This makes it hard to get the more complicated, interactionist interpretation of events across. Some professionals then conclude that 'it's better to keep quiet about things'. Furthermore, throughout the research, it has been shown that the anecdotal, vulnerable, personal, and 'one valley' knowledge of residents (Scott 1998, p. 317), in discussions with e.g. housing company officials, is often rejected as merely casuistry: unfortunate individual incidents, without relevance to general policy.

8.7 Conclusion

As an interstitial space, the Rabot atelier offered the practical and mental space for residents to stage identities on the threshold of their formal and informal position in social housing. In this way, the communal atelier functioned as a symbolic repair workshop (Hillaert 2014; Sennett 2009), renegotiating between the inside and outside world and deciphering the different encoding of both. For a brief moment, the workspace offered a symbolic façade, within the inside world as well as towards the outside world, and therefore, the possibility to stage new social identities through which meaning and authorship of narratives could arise.

Interstitial spaces and façade regions are necessary conditions for social housing tenants to escape the reproduction of stigma and the social closure of their living environment. Only then is there a possible co-creation of roles and knowledge between the inside and the outside world of these buildings, thereby valuing the experience of all tenants—capable or otherwise.

References

Allemeersch S (2022a) The decline of the Rabot towers: Prak & Priemus (1986) revisited [Manuscript submitted for publication]

Allemeersch S (2022b) Housing pathways of social high-rise residents in the Rabot towers in Ghent: From 'Old Belgians' to the new marginality [Manuscript submitted for publication]

Allemeersch S, Capelle B, Debaene M (Eds).(2014) Rabot 4–358 (n.p.)

Blokland T (2019) 'We live like prisoners in a camp': Surveillance, governance and agency in a US housing project. In: Flint J, Powell R (eds) Class, ethnicity and state in the polarized metropolis: putting Wacquant to work. Palgrave Macmillan, pp 53–80

D'Cruz H, Jones M (2004) Social work research: ethical and political contexts. Sage

De Decker P, Pannecoucke I (2004) The creation of the incapable social tenants in Flanders, Belgium: an appraisal. J Housing Built Environ 19(3):293–309

Dehaene M (2008) Porosity: Jean Remy and the ecology of weak integration [Unpublished manuscript]

Dehaene M (2011) Jean Remy and the urban ecology of weak integration [Unpublished manuscript]

Garneau S, Namian D (Eds) (2017) Erving Goffman et le travail social. University of Ottawa Press

Goetz E (2011) Where have all the towers gone? The dismantling of public housing in U.S. cities. J Urban Affairs 33(3):267–287

Goffman E (1966) Behaviour in public places: notes on the social organization of gatherings. The Free Press. (Original work published in 1963)

Goffman E (1996) Gestichten: De sociale situatie van gedetineerden, psychiatrische patiënten en andere bewoners van 'totale inrichtingen' (Nijhoff P, Johannes GJ, Trans.). Bijleveld. (Original work published 1963)

Goffman E (2004) Frame analysis: an essay on the organization of experience (Berger BM, Ed.). Northeastern University Press. (Original work published in 1974)

Goffman E (2018) Stigma: Notities over de omgang met een geschonden identiteit (Johannes GJ, Mordegaai L, Trans.). Bijleveld. (Original work published in 1963)

Goffman E (2019) De dramaturgie van het dagelijks leven: Schijn en werkelijkheid in sociale interacties (Nijhoff P, Johannes GJ, Trans.; J. Mordegaai, Ed.). Bijleveld. (Original work published in 1959)

Harvey D (1973) Social justice and the city. University of Georgia Press.

Hillaert W (2014) In het midden aan de rand. In Allemeersch S, Capelle B, Debaene M (Eds) Rabot 4–358, pp 149–155. (n.p.)

Jacobs J (1961) The death and life of great American cities. Modern Library ed. Modern Library

Lipsky M (1980) Street level bureaucracy: dilemmas of the individual in public services. Russell Sage Foundation

Luyten S (2014) Conciërge Rabottoren steelt 20.000 euro van invalide bewoonster. Het Nieuwsblad. https://www.nieuwsblad.be/cnt/blslu_01234980

Rancière J (2010) Dissensus: on politics and aesthetics (Corcoran S, Trans.). London: Continuum

Remy J (1993) Trends in urban sociology in French-speaking countries from 1945 to 1980. Geo J 31(3):265–278

Remy J, Voyé L (1981) Ville, ordre et violence : formes spatiales et transaction sociale. Presses Universitaires de France

Rondelez E, Bracke S, Vandekinderen C, Roets G, Bracke P (2018) Revisiting Goffman: frames of mental health in the interactions of mental healthcare professionals with diasporic Muslims. Soc Theory Health 16(4):396–413
Sahlin I (1996) From deficient planning to 'incapable tenants'. Changing discourses on housing problems in Sweden. Scandinavian Housing Plann Res 13:167–181
Scott JC (1987) Weapons of the weak: everyday forms of peasant resistance. Yale University Press
Scott JC (1990) Domination and the arts of resistance: hidden transcripts. Yale University Press
Scott, J. C. (1998). Seeing like a state: how certain schemes to improve the human condition have failed. Yale University Press.
Sennett R (1970) The uses of disorder: personal identity & city life. Knopf
Sennett R (1978) The fall of public man. Random House
Sennett R (2009) The craftsman. Penguin
Sennett R (2018) Building and dwelling: ethics for the city. Allen Lane
Vervloesem ES (2019) Stedenbouw en sociale verandering: Leren van vergeten geschiedenissen van stad maken. Technische Universiteit Eindhoven
Wacquant LJD (2004) Body and soul: ethnographic notebooks of an apprentice boxer. Oxford University Press
Wacquant LJD (2008) Urban outcasts. Polity Press, A comparative sociology of advanced marginality
Wacquant LJD (2019) Class, ethnicity and state in the making of urban marginality. In: Flint J, Powell R (eds) Class, ethnicity and state in the polarized metropolis: Putting Wacquant to work. Palgrave Macmillan
Warming H, Fahnøe K (Eds) (2017) Lived citizenship on the edge of society: rights, belonging, intimate life and spatiality. Palgrave Politics of Identity and Citizenship Series

Chapter 9
From a Community of Practice to a Community of Planning: The Case of the *Sansheroes* Network in the San Siro Neighbourhood in Milan

Elena Maranghi

Abstract In the current framework of welfare shrinking, it is highly necessary to transform citizens and local organizations from targets into co-producers of urban policies. Moreover, even though large-scale social housing estates are often characterized by social exclusion and high levels of socio-economic vulnerabilities, they at the same time represent 'local tanks' of competencies and social resources. In these regards, the 'empowering planning' approach—referring to the valourization of local competences and expertise within urban regeneration processes—has positive impacts, both in terms of socio-economic inclusion and the 'expansion' of active citizenship among local actors and in terms of designing more effective policies, enriched by local perspectives and know-how. Based on the analysis of a pilot action developed within the SoHoLab project in the San Siro neighbourhood that fostered the empowerment of a local grassroots network, the chapter examines processes of recognition and reinforcement and the promotion of local competencies, outlining their different phases and the characteristics of the groups involved. It will highlight the transition from a community of practice to a community of planning that is able to develop visions and actions aimed at a shared regeneration of a certain area.

Keywords Local network · Community of practice · Knowledge coproduction

9.1 Introduction

The field of urban studies is increasingly characterized by an open and hybrid approach that involves the cross-fertilization of 'expert' and 'locally produced' knowledge. The Urban Living Lab methodology constitutes a relevant part of this tendency, as it includes different and multiple stakeholders in the analysis of a certain problem and the design of shared solutions (see Cognetti, this volume). Indeed, in some way, the complexity of contemporary cities is 'asking us' (Cognetti and Fava 2017)—as researchers and practitioners in the fields of urban studies and

E. Maranghi (✉)
Milan, Italy
e-mail: elena.maranghi@gmail.com

N. Aernouts et al. (eds.), *Urban Living Lab for Local Regeneration*,
The Urban Book Series, https://doi.org/10.1007/978-3-031-19748-2_9

urban planning—to go beyond our disciplinary positions, as such complexity needs more adequate categories in order to be interpreted correctly. Our perspectives—as architects, planners, anthropologists, geographers, sociologists, etc.—need to be 'contaminated' by one another in order to get closer to a deeper understanding of ordinary life in our cities (Cellamare 2016; Cognetti and Fava 2017). As Cognetti and Fava (2017) have extensively described, it can be stated that a city is to a certain extent an object that cannot be disciplined and that has a character of resistance: 'it resists being reduced to a disciplinary order and always emerges as the extraordinary that forces the single disciplines to think from the outside' (Cognetti and Fava 2017, p. 134).[1] If, on the one hand, this aspect invites different disciplines to question themselves about their ability to open up to other perspectives, interpretative frameworks, and tools, on the other hand, it also pushes them to be open to the contribution of 'nonexpert' knowledge. Indeed, over the last decades, urban areas have become increasingly complex arenas of actors. Many different stakeholders are today recognized as legitimate 'carriers' of urban knowledge, which becomes essential for promoting more effective knowledge frameworks and public policies.

Therefore, contemporary cities challenge us to go beyond interdisciplinarity or, from another point of view, to turn it into its more profound meaning: the capacity to listen to and collect knowledge that arises from urban space itself through relational observation. This kind of observation considers 'local knowledge, knowledge of everyday life, a knowledge that is embedded in the actions and reflections of the subjects engaged with. This means to have an interdisciplinary but also non-disciplinary way of thinking and producing knowledge' (Pizzo and Rolnik 2019, p. 30). This attitude is fundamental in both a problem-setting and a problem-solving perspective. Indeed, interdisciplinarity questions the effectiveness of the languages and tools used to interact[2] with urban areas and propose or produce any change.

If, in an initial phase, this trend found an outlet in the promotion of participatory approaches, in recent years we have moved from a so-called phase of 'weak participation' (Cognetti 2007) to an ever greater role for those who, for various reasons, inhabit places (residents, local activists, third-sector actors rooted in certain spatial contexts, etc.), especially in some contexts within the Italian panorama. Local actors have increasingly become 'experts' in their own spatial field and started to participate in a process of producing social and 'enlarged' knowledge (Ranzini 2018). Especially in more challenging contexts,[3] their activation becomes fundamental in any process of policy initiation and implementation. The structural lack of knowledge that characterizes large-scale social housing estates[4] pushes engaged researchers and practitioners to question themselves about how knowledge is produced and shared

[1] Translated by the author.

[2] To describe them, to promote participation within them, etc.

[3] Such as peripheral and marginalized territories.

[4] I refer here especially to the fact that, in the experience of the SoHoLab case studies, housing associations in large-scale social housing estates do not usually have a clear picture of the dwelling conditions and condition of the built stock, particularly in neighbourhoods where informal dwelling and squatting are more widespread.

in these places. This regards not only the evaluation of the meanings and targets of knowledge production but also how and which sources of knowledge are selected, which materials are collected and elaborated on, etc.

It is particularly important to underline how 'nonexpert' knowledge plays a crucial role in these contexts, as it is characterized by significant informality or, in some cases, even by 'parallel' systems of values. Lave and Wenger call this phenomenon 'situated learning' (2010): learning that takes place through the relationships between people and through connecting prior knowledge with authentic, informal, and often unintended contextual learning. By relating to these different sources of knowledge and through the process of situating (see Cognetti, this volume), researchers and practitioners acquire a 'more local' perspective, which anthropologists would define as an 'emic perspective' (Vereni 2015).

Based on these assumptions and the action research developed by the Mapping San Siro group[5] within the SoHoLab project in Milan, the contribution discusses the role that the activation of local networks through the coproduction of knowledge can assume in large-scale social housing estates. It analyses the experience of the *Sansheroes* network, which was developed in the San Siro social housing neighbourhood.[6]

9.2 Mapping San Siro as a Multisource Observatory

As stated above, within the SoHoLab project framework, interdisciplinarity was conceived not only as a profound collaboration among different disciplines[7] but also as the combination of 'expert' and locally produced knowledge. During the entire development of the Urban Living Lab implementation, this knowledge assumed multiple forms and meanings.[8] To produce local knowledge meant, then, not only to 'include' dwellers but also to stimulate collaboration and exchange among the

[5] Mapping San Siro is an action-research group belonging to the Department of Architecture and Urban Studies of the Politecnico di Milano and is part of the Polisocial programme. Coordinated by Francesca Cognetti, with the support of Liliana Padovani, it has been active since 2013. In 2014, it opened a space in the San Siro district. From 2017 to 2020, Mapping San Siro was part of the SoHoLab project (see the Introduction of this book).

[6] With its approximately 6,000 dwellings and 12,000 residents, the San Siro district is one of the largest public housing complexes in the city of Milan. Built between the 1930s and 1950s, the district is today located in a semi-central position with respect to the urban fabric, but it is characterized by dynamics of social exclusion and economic deprivation. San Siro is one of the most multicultural districts in the city: about 50% of the resident population is of non-Italian origins. The neighbourhood suffers from severe decay of most buildings and is characterized by the presence of many empty apartments, an element that has contributed to the consolidation of squatting practices.

[7] Represented in the composition of local research teams.

[8] This aspect is clearly linked to the concept of 'situating' (see Cognetti and Grassi, this volume), which enabled the research group to share intensive and long-term 'engaged ethnography' (see Aernouts et al., this volume) in the field, during which it was able to constantly gather knowledge from local actors.

different professionals and activists who were working in the considered urban area,[9] whose contribution appeared to be crucial in the different development phases of the SoHoLab Urban Living Lab: from the understanding the local context, to the implementation of the pilot projects developed, to the evaluation of the research and the actions promoted.

From the very beginning of its research activities, and in line with what was argued at the beginning, the position of the Mapping San Siro action-research group was characterized by the organization of dialogue and the creation of a relationship with the various players present in the area. This approach was seen as a necessity to face the almost 'annoying complexity' (Scanni 2015) present in the context, with respect to which it was essential for the research to be based on a solid local mediation. What can be defined as the neighbourhood's opacity regarding fact-finding, pushed the group towards this orientation as well: there was a lack of reliable data and clear sources to refer to as well as the presence of numerous informal and complex dynamics that were objectively difficult to investigate without a grounded and situated knowledge of the context.

This opacity fed the stereotypical and stigmatizing image of San Siro, with the effect of not only consolidating the negative perception of the neighbourhood by 'the outside' but also determining the worsening of internal conflicts and of the difficulty of dealing with local problems. These problems were often too unclear or seemed insurmountable: as can be guessed, it also inhibited the capacity for action of the various actors present in the neighbourhood. Often, in their relationship with Mapping San Siro, a university group, these actors demanded more public knowledge to emerge and to be shared (Cognetti 2018a). Since the group did not have a clear and defined 'client' at the time (see Castelnuovo and Cognetti 2019), local subjects—who expressed requests that were certainly fragmented but also concrete and situated in relation to the possible roles that research could assume—progressively became not only the most direct interlocutors and 'informants' of the group but also the main 'clients' on Mapping San Siro's research path. Therefore, one of the main aims of the research, from the very beginning, was to systematize and collect knowledge on the neighbourhood in order to build a 'multisource observatory' in which different forms and types of knowledge had the possibility to converge and meet: local knowledge, linked to inhabiting the place or based on professional practice (third-sector organizations) or even on activism (voluntary associations, informal groups); quantitative and qualitative sources produced by institutions; research of an anthropological and socio-spatial nature, produced by the university and independent researchers, etc.

From this perspective, based on an interdisciplinary approach, Mapping San Siro tried to push itself to interpret interdisciplinarity as an interweaving of knowledge (and not just disciplines) of different natures. The enhancement of common knowledge (Dewey 1938) produced by so-called everyday-makers (Bang and Sørensen 1999)—who are experts in a certain context in a different way (Cognetti 2018b)—thus became a full part of the research, helping the group to access and interpret a

[9] Social workers, activists, local administrators, etc., who usually acquire knowledge through experience and practice (see Dewey 1938).

context which was very difficult to permeate. The main objective was to construct a knowledge base that could become the engine for the improvement of local change and development. In this sense, the group was also reflecting on the role of knowledge coproduction for local organizations, imagining it as an accessible and usable asset for improving and strengthening the capacity of each actor to act. In order to act in this direction, however, the first step the group needed to take was to recognize local knowledge as valuable. Second, the group had to help local organizations and activists to structure this knowledge through moments of self-reflection.

9.3 Towards a Competent and Structured Local Network

Following the consolidation of the process of gathering local knowledge, also thanks to the establishment of the research group headquarters in the neighbourhood,[10] it became progressively more evident how the strong internal fragmentation among different groups and activists represented a problem with regard to the possibility of influencing public discourse and having a voice in the structuring of urban policies aimed at the neighbourhood. Indeed, internal fragmentation and latent conflict among locals seemed to play an important role in a certain dynamic of exclusion of the neighbourhood from the framework of city policies. It appeared to limit the capacity for implementing collective action and improving local agency. Especially in the context of Milan, a city characterized by a strong, proactive role of the third sector and a high level of civic participation in the definition of policy frameworks, the 'ability' of local activation (Ranzini 2018) strongly influences the possibility of attracting public investment and financial resources to different urban areas.

Therefore, creating a structured path of co-construction and of sharing local knowledge related to the neighbourhood and its conditions became one of the priorities of the Mapping San Siro research group, as this was seen as a tool to open negotiations and foster a dialogue with the institutional level. A first step in this direction took place in November 2016, when around fifteen representatives of local associations and cooperatives as well as activists accepted an invitation to meet at the Mapping San Siro headquarters in order to discuss the situation of the neighbourhood. The intention of the research group was to use the perspective and knowledge of each of the participants (and their respective organizations) in order to produce a shared, adequate, and updated vision of the socio-spatial scenario for San Siro. This was intended as a tool capable of, on the one hand, guiding local organizations in dealing with the complexity of the neighbourhood and, on the other—once the scenario was transferred to the competent institutions—calling the institutional level to counteract the critical conditions the neighbourhood was suffering. These included phenomena such as educational poverty, social isolation and exclusion, and

[10] In 2014, Mapping San Siro obtained a free concession from Aler Milano to reopen a former shop in the neighbourhood, located at the street level, to be used for research activities. Afterwards, in 2019, Politecnico of Milan obtained a larger space, located a short distance away from the first one.

the bad maintenance of public spaces. About ten organizations were represented at the table: both more formal and structured entities and organizations of a voluntary and informal nature.

The existence of a trust relationship between Mapping San Siro and the represented organizations was key to stimulating their willingness to participate. In fact, between 2014 and 2016, Mapping San Siro activated some pilot projects in collaboration with different local entities, gradually consolidating and deepening everyday relationships in the field.

Another relevant factor that pushed the meeting was the continued exclusion of the neighbourhood from several municipal programmes and policies directed at urban peripheries. This generated a proactive reaction in those who were engaged in improving the neighbourhood's conditions on a daily basis. It was the beginning of a slow but constant process involving the construction, expansion, and consolidation of a local network, which I will try to reconstruct as five different stages below (Maranghi 2019).

9.3.1 Getting to Know San Siro Together (November 2016–December 2017)

Through five meetings and a subsequent re-elaboration of the content that had emerged, the local actors who gathered around the table produced a first 'snapshot' of the current state of the neighbourhood. This document aimed to bring out the latent knowledge capital of the various organizations and to dissect the emerging issues related to the neighbourhood through the recognition of problems and resources. The document was built on the knowledge gathered by each organization in its daily practice and also included a series of initial considerations related to possible actions to promote change. The result of this phase was the publication of an online, open, and accessible document (*Fotografia del quartiere*[11]). After the first meetings, the group decided to establish itself as a network of organizations and people, which was given the name of *Sansheroes*.[12] It also decided to set up a mailing list and monthly meetings for mutual updating and collaboration.

9.3.2 Sharing the Path with the Neighbourhood (January 2018–June 2018)

As part of the local network and in collaboration with it, Mapping San Siro, organized a series of focus groups and dialogues aimed at sharing the document mentioned above with target groups that were representative of some relevant populations in the

[11] https://issuu.com/52340/docs/fotografia_del_quartiere_san_siro_2.

[12] The play on words refers to the name of the San Siro district: it sounds like 'heroes of San Siro'.

neighbourhood (the elderly, young people, women of foreign origin, etc.). Observations, suggestions, and criticisms were collected in order to enrich and consolidate the document. The objective of this phase was to validate, together with residents, what had been elaborated by the network. The result of this second phase was an expanded and revised version of the document. During this period, the network further expanded itself, welcoming new organizations from the neighbourhood.

9.3.3 Designing Together (June 2018–December 2018)

The members of the local network worked together to design a vision for the future of the San Siro neighbourhood, outlining five possible areas of intervention to be strengthened and developed: interculturality, education, local competencies, housing policies and support for vulnerabilities, and the quality of the built environment. Some first hypotheses of strategic lines of action were developed and were associated with the five areas. The document was then expanded to include this design evolution, enriched and completed by some maps drawn up by Mapping San Siro on the basis of collective work. A new version entitled *Istantanee di San Siro. Presente e future del quartiere* was then published online.[13]

9.3.4 Involving the City (January 2019–February 2019)

The local network stimulated the sharing of the process and its results with institutions and with the city in order to show the potential of the path and at the same time call on the responsible institutions to implement actions to improve the living conditions in the neighbourhood. Therefore, it was decided to organize a city-level public presentation of the document and of the network itself.[14] Representatives of the main local institutions and civil society were invited to the event, which took place in February 2019 and was open to the public. Although the institutions publicly expressed an important signal with their participation and with the declaration of their intention to work together with the network, this did not happen in the following weeks.

[13] https://issuu.com/52340/docs/istantanee_2019.

[14] This took place at the Triennale of Milan, one of the most relevant cultural institutions of the city.

9.3.5 Dialogue with Institutions and Consolidating a Planning Vision (March 2019–February 2020)

In April 2019, the Lombardy Region and the Municipality of Milan, in the presence of the Aler Milano[15] Housing Agency, decided to meet at the Mapping San Siro headquarters in the neighbourhood, declaring their willingness to work together in the neighbourhood on an experimental basis. However, this position was once more not followed by further steps. The network, therefore, decided to set up some internal working groups to continue implementing concrete proposals and strategies. The chosen themes were: housing and living conditions; public art; job opportunities; and public space, courtyards, and waste management. The network members also continued to work together to develop applications for local calls for tenders and initiatives. They were able to win some funds and bring new energies to the neighbourhood. Moreover, the network continued to expand itself through the participation of other organizations active in the neighbourhood and also attracted new professional competences and city-level organizations.

9.4 From a Community of Practice to a Community of Planning

A first key aspect to underline in the reconstruction of the above-described path concerns the possibility for local actors to recognize themselves as a collective subject, which could be defined as a community of practices (Wenger 1998, 2010) that is intended, first of all, as a co-learning context. In this case, the practice that generated and reinforced the *Sansheroes* network can be identified with the local production of welfare, both at a formal and an informal level. In this sense, the dynamics of the transformation of welfare, which rooted it in an increasingly local perspective (Vitale 2007), made it possible for some professional figures (but not necessarily strictly professional ones) to tie themselves to certain specific areas. They built cognitive expertise and relational capital, which transformed them into increasingly competent actors in relation to the dynamics of the context's urban and social regeneration, beyond their specific area or sector of intervention. Just to mention some of the most significant examples: the school parent committee acquired skills in analysing and collecting social needs of families that had recently arrived in the neighbourhood, mostly from other countries, and consequently implemented small mutual aid networks; volunteer women teaching Italian to foreigners became the most 'expert' in understanding the needs of the female population of foreign origin, to which they dedicated specific courses; and so on. It appears, then, that the field of

[15] The Regional Public Housing Agency which owns and manages the public housing stock in the neighbourhood.

'territorial welfare' is configured as a field of shared and mutual learning, which the different local actors feel they belong to. They are, at the same time and to different degrees, both competent subjects and learners.

While these processes are increasingly widespread and significant in the contemporary city, they simultaneously require considerable investment of resources and intentions. This usually corresponds to a first step of self-recognition and mutual recognition among local actors, of competencies, knowledge, and skills which are significant not only in daily practice but also in a broader and more systematic perspective. A first element at the basis of the constitution of the *Sansheroes* network was, therefore, precisely the construction—promoted, in the first place, by Mapping San Siro—of a fertile context for the recognition of the fundamental role that locally produced knowledge and action[16] play in policy arenas. By locally produced knowledge, I refer to knowledge which is the expression of various local actors and different perspectives and which, above all, does not necessarily take the form of exclusively scientific and expert or exclusively local knowledge. Instead, it assumes a hybrid character, and therefore, becomes more effective as it configures itself as a bridge between social worlds, skills, and organizations.

In this sense, the work with the *Sansheroes* network started at first as a process of local actors' self-recognition, identification, and mutual recognition of the 'dignity', the value, and the crucial role played by spatial competences and locally produced knowledge. The three aspects (self-recognition, identification, and mutual recognition) of the process allowed local actors to become familiar with their competences and endorse the legitimacy of their voices (self-recognition); thereby understanding the relevance of their work and their points of view for the policy arena (identification); these aspects, as was the case for *Sansheroes*, can then lead to the breaking of a spiral of local fragmentation, stimulating local organizations and activists to acknowledge each other's role (mutual recognition) and converting conflict from a barrier into a treatable problem (Padovani 2016). Naturally, this process does not have an immediate effect, as it operates as a long, incremental, and nonlinear negotiation. This process needs a kind of 'mediator', as was the case with Mapping San Siro: the research group played a bridging and facilitating role through an open approach and in virtue of a certain degree of impartiality which characterizes the university.[17]

It is also important to underline the characteristics and nature of the knowledge produced. In the above-described case, Mapping San Siro implemented from the

---

[16] I refer here to the concept of 'situated learning' (Lave and Wenger 1991), which underlines the relational character of knowledge, constantly negotiated in different contexts (Cognetti 2018b). It is no coincidence that as Mapping San Siro, we refer to the concept of situating as a practice of building knowledge carried out by the research group through a deep and slow rooting in the context, based on the multiple relationships with local interlocutors and places we engaged in.

[17] From a methodological point of view, it should be underlined that 'in this process of building knowledge it is important to gild oneself with tools that give space to the contributions that will emerge from common knowledge and from situated knowledge (interviews, dialogues, informal contacts, work tables, participant observation, etc.)' (Cognetti 2018a., p. 35); just as it is essential to activate skills linked to the relational and so-called affective dimension of planning (Kondo 2012), capable of leveraging the degree of rootedness in the local context and civil commitment of each subject involved.

very beginning an action-research approach in an attempt to counteract the inertia of the local context and the paralyzing effect of the public image of the neighbourhood as a unitary object that could not be discussed or changed (Cognetti 2018a). From the very start—originally on an extremely small scale, then gradually extending more significantly—Mapping San Siro developed several pilot projects aimed at testing and stimulating change (see Cognetti et al., this volume). These projects were aimed at encouraging, guiding, and supporting local action, gathering and elaborating usable knowledge on certain matters (such as public space, housing, educational poverty, etc.) in order to strengthen and to some extent further legitimize local subjects in their action.

In this process, therefore, the collection of knowledge is profoundly linked to the theme of action: in the action-research paradigm, indeed, change occurs when action and discourse become complementary. This happens when the actors involved in a common path carry out a concrete action together (Barbier 2007): when knowledge structures and stimulates the action and thus produces new knowledge. In the case of *Sansheroes*, several concrete actions can be identified which, as they were designed together, contributed to the self-recognition of the network: the coproduction of the above-mentioned documents, their publication online, the public presentation (to the neighbourhood and the city), the co-design of pilot projects, etc. In this process, the co-construction of shared knowledge through action functioned as an 'enzyme' for the cohesion of the network, as co-designed actions really represented a shared field, accessible to every actor who participated in the process. Moreover, recognizing itself as a competent community (Iscoe 1974) stimulated the *Sansheroes* network to become a permanent planning platform for and with the neighbourhood. This aspect implied a significant transfer and activation of design skills and spatial sensitivity from the Mapping San Siro group to the other entities of the network, linked to the ability to promote complex visions of transformation and to activate co-design groups. This transition can be described as the transformation from a community of practice into a community of planning, able to project itself and the neighbourhood in a vision of local, complex, and integrated regeneration which goes beyond the specific fields of action represented by the different actors involved.

It is interesting to note, in this regard, how a 'peripheral actor', which the Mapping San Siro group was at the beginning of its activity in the neighbourhood, acts with respect to a certain learning context. On the one hand, through what Lave and Wenger (2010) defined as apprenticeship, peripheral actors are able to recognize and reveal the cognitive and symbolic expertise associated with a certain community of practice, precisely by virtue of occupying a position on the sidelines. On the other hand, as initially 'external' subjects, they could bring new skills that, in certain cases, prove to be essential for consolidating and strengthening the community of practice. In the *Sansheroes* case, these are the skills associated with urban analysis, spatial codesign, planning vision, etc.

In conclusion, it can be stated that, even if the path described here did not naturally produce a definitive solution to existing problems associated with the neighbourhood, as the network members continued to work together even after the end of the SoHoLab project, the emergence of the network represented a fundamental advancement in the recognition of the neighbourhood in policy arenas and in the capacity of local actors to produce more effective, coordinated, and integrated interventions and projects. It appeared to be an especially valuable aspect, for instance, in facing the first stages of the pandemic crisis in 2020, which strongly hit more fragile areas of our cities, such as San Siro.

References

Bang HP, Sørensen E (1999) Everyday maker: a new challenge to democratic governance. Administr Theory Praxis 21(3):325–341

Barbier R (2007) La ricerca-azione. Armando

Castelnuovo I, Cognetti F (2019) Mapping San Siro Lab: experimenting grounded, interactive and mutual learning for inclusive cities. Trans Assoc Europ Schools Plann 3:37–54

Cellamare C (2016) Leggere l'abitare attraverso l'interdisciplinarietà e la ricerca-azione. Territorio 78:28–39

Cognetti F (2007) Parlare di partecipazione. In: Milano G (Ed), La partecipazione in provincia di Milano. Ricerche e indagini per una interpretazione del territorio. Provincia di Milano, pp 162–169

Cognetti F (2018a) San Siro come campo di apprendimento. Diario di bordo verso un think thank di quartiere. In: Cognetti F, Padovani L, Perché (ancora) i quartieri pubblici. Un laboratorio di politiche per la casa. Franco Angeli, pp 15–31

Cognetti F (2018b) Apprendere la città. Le prospettive di una ricerca inclusiva e relazionale. In: Cognetti F, Padovani L, Perché (ancora) i quartieri pubblici. Un laboratorio di politiche per la casa. Franco Angeli, pp 33–51

Cognetti F, Fava F (2017) La città indisciplinata. Note per un'agenda di ricerca. Tracce Urbane 1:126–136

Dewey J (1938) Experience and education. Macmillan

Iscoe I (1974) Community psychology and the competent community. Am Psychol 29(8):607–613

Kondo MC (2012) An ethic of love for planning. In: Porter L, Sandercock L, Umemoto K, Bates LK, Zapata MA, Kondo MC, Zitcer A, Lake RW, Fonza A, Sletto B, Erfan A, Sandercock L (eds), What's love got to do with it? Illuminations on loving attachment in planning. Plann Theory Pract 13(4):593–627

Lave J, Wenger E (1991) Situated learning. Cambridge University Press, Legitimate peripheral participation

Maranghi E (2019) Il ruolo della conoscenza locale nell'attivazione di comunità di pratiche e di progetto. Il caso della rete Sansheroes (Milano). Tracce Urbane. Rivista Italiana Transdisciplinare Di Studi Urbani 3(6):177–195

Padovani L (2016) Partecipazione come forma di apprendimento. Sentieri Urbani 7(21):39–43

Pizzo B, Rolnik R (2019) On interdisciplinarity and non-disciplinary research: a conversation. Tracce Urbane. Rivista Italiana Transdisciplinare Di Studi Urbani 3(6):23–34

Ranzini A (2018) Troppo competenti? Riflessioni sull'abilitazione delle competenze degli attori locali nei processi di rigenerazione urbana di contesti marginali. Tracce Urbane 3:115–133

Scanni M (2015) San Siro il quartiere laboratorio. Magzine. Periodico della scuola di giornalismo dell'università Cattolica. Online. http://www.magzine.it/wp-content/uploads/2015/05/magzine_19_-SanSiroStories.pdf

Vereni P (2015) Addomesticare il welfare dal basso. Prospettive e paradossi delle occupazioni abitative romane. Meridiana 83:147–169

Vitale T (2007) Integrazione e territorializzazione del welfare. Alcune implicazioni per i servizi sociali. Voci di strada 19(3):91–113
Wenger E (1998) Communities of practice: learning, meaning and identity. Cambridge University Press
Wenger E (2010) Communities of practice and social learning systems: the career of a concept. In: Blackmore C (ed) Social learning systems and communities of practice. Springer, pp 179–198

Part III
Approaching Space in Large–Scale Social Housing Estates

Chapter 10
Marginalization Through Mobility and Porosity: How Social Housing Dwellers See and Live the City

Lucia Capanema-Alvares

Abstract Urban mobility in its broader meaning has become fundamental in neoliberal times, for it determines who gets what, how often, and at what cost. While motility is a component of mobility—together with connectivity and reversibility—defined by Kaufmann (2014) as a quality of the actor and/or of the dialectical relation between the self and the field of the possible, and accessibility concerns the structures necessary to take part in this possible, porosity is a quality of the territory and/or of the dialectical relation between space and society. The three of them inseparably carry the city dwellers' possibilities of fulfilling their projects and wishes in the city territory. In order to start picturing how the society–space dialectic based on motility, accessibility, and porosity shapes daily social relations, especially where spatial justice is at stake, this study—part of the all-encompassing Action Research Project Mapping San Siro—surveyed 100 inhabitants of the Milanese neighbourhood. The resulting picture is a snapshot, a working scenario which helps bottom-up initiatives understand and focus on the most problematic, sometimes underlying aspects of marginality. While the quantitative results point to low-income inhabitants who work hard, use public transportation on an everyday basis, have few, if any, professional dreams, and feel reasonably welcome in a city they did not choose to live in, a number of qualitative results show that mobility (as a whole social phenomenon) problems can be deeper, not yet surfaced or voiced.

Keywords Mobility · Motility · Accessibility · Porosity · Spatial justice · Marginalization

10.1 Introduction

Historically, the distribution of services, facilities, amenities, and open public spaces in the city has been a social construction, unequal and responsive to capital accumulation needs. As a result of 'governance' rescaling, policies' reorientation towards

L. Capanema-Alvares (✉)
Universidade Federal Fluminense, Niterói, Brazil
e-mail: luciacapanema@gmail.com

N. Aernouts et al. (eds.), *Urban Living Lab for Local Regeneration*,
The Urban Book Series, https://doi.org/10.1007/978-3-031-19748-2_10

competition, and the reduction of states, this has worsened since the 1980s (Purcell 2002). Hence, 'mobility' in its broader meaning becomes increasingly important: it determines who gets what in the urban milieu, how often, and at what cost, in a scenario marked by accumulation by dispossession (Harvey 2004).

City fluxes are at the basis of how *citadins* perceive, conceive, and live the environment and how they ontologically relate to the city and the Other. Through momentary or lasting appropriations of public spaces, city dwellers entail a social dialectic that is not only established in space, but also undertaken with space (Santos 1996).[1] Underlying this process is Santos' phenomenological notion of space as a whole, as a social instance as well as economic, cultural, ideological, and political instances. The society–space dialectic is based on motility, accessibility, and porosity as much as on concrete urban spaces and structures; these together will shape the daily social relations, cognitions, and expectations of all city inhabitants.

Uncovering mobility influences in social relations is one of the first steps towards the recognition and valorisation of *citadins*' experiences and their demands for a city of rights and of 'the right to produce urban space that meets the needs of inhabitants' (Purcell 2002, p. 103). A city in which all *citadins* can come and go safely, are recognized, and can feel welcome to imagine, pursue, and nurture their projects, wishes, and desires in the whole territory: a more mobile, accessible, and porous city for all, socially constructed by all of its dwellers.

The research results herein discussed stem from the eight-year-long action research project 'Mapping San Siro', one of the most important academic community efforts to translate what segregated and multi-ethnic neighbourhoods have to tell us in contemporary Europe. This authentic 'living lab' praxis concerning housing, public spaces, education, and women has welcomed individual and collective projects such as my 'Mobility as key towards sustainable and just cities in the twenty-first century' through an international research position at the Department of Architecture and Urban Studies, Politecnico di Milano.

10.2 The Distribution of Services, Facilities, Amenities, and Open Public Spaces Among Different Agents of the City

As the growth and distribution of the world population and of economic benefits expand unequally, all large metropolises in which diverse populations live will, in a few years, be affected by the same socio-economic and environmental problems to varying degrees, according to their social, cultural, economic, and institutional morphologies; these morphologies define the different levels of rights to the city and, therefore, of social inequality (Secchi 2011). The opportunities offered to individuals according to their ability to aggregate/explore the various places in the city, i.e.,

[1] Many terms and quotes presented in this chapter were freely translated by the author, as they have not been translated and published in English.

their space capital (Soja 2010), become an autonomous, central, and independent variable in understanding Secchi's 'new urban issue'. In other words, since urban mobility, accessibility, and porosity regulate the access of the rich to the city and/or the incarceration of the poor in it (Secchi 2011), they become some of the main determinants of social inequity.

Bourdieu (1993), concerned with the unequal distribution of symbolic and cultural capital arranged in social space, uses the idea of a force field in which agents, disposed by their quantity of capital, move in a certain way according to the movements of the antagonistic agents. The distribution of capital and the organization of agents in social space can constitute identification processes that establish the collective possibility of conflicts and ruptures, deconstructing and recreating the disposition of social space (Bourdieu 1996). In this case, conflicts would constitute a porous process, activating relationality rather than separation, as Stavrides (2018) expects and will be further discussed below. On the contrary, similar situations allow, under Bourdieu's logic, for a spatial analogy between the dominated and the dominant and can produce an effect of position homology in the dominated agents—creating subserviency and mimicking.

Davis (1993), Barbosa and Dias (2013), Fernández-Álvarez (2017), and Tângari (2018), among others, have shown the historically unequal distribution of facilities, cultural equipment, and green and public spaces around the world. While Tângari (2018) specifically sees the political-administrative dimension as a dominant force in the production of unequal public spaces to the detriment of sociability and encounters, Sheller (2018b) calls our attention to public spaces as social instances with the potential to house and support commonalities, where marginalized groups may articulate the under-commons (i.e. migrants, artists, street-vendors, and all of those who usually share difficulties and resources in public spaces) in spite of the harassments to which they are subjected.

10.3 Motility, Accessibility, and Porosity: Forms and Fluxes Shaping Social Relations

The way people live spaces as a dialectical result of their perceptions and cognitions is based both on the 'brute facts' (Cresswell 2006) of our socially constructed urban forms and on people's access (through mobility and porosity) to these forms, as argued in the previous section. What we seek to research, comprehend, document, and showcase is how the society–space dialectic based on motility, accessibility, and porosity shapes daily social relations, cognitions, and expectations. In Bourdieu's habitus—a system based on 'visions and division constitutive of a social order' (Bourdieu 2000, p. 143)—objective possibilities conform imaginaries and subjective expectations; expectations are conditioned by the subjects' perceptions of their own social position. Personal identity and social position identification would come from each individual's complex and multiple representations/imaginaries of reality, given each one's position in the urban habitus (Bourdieu 2000). Bourdieu also understands habitus as a result of class trajectories, as he makes a clear distinction between middle classes' habitus—based on freedom of consumption—and working classes' habitus, based on consumption needs. When Bourdieu establishes a contrast between middle classes' and working classes' habitus, he also shows that capital forges a system of perceived differences dependent on the habitus and on objective material conditions (Bourdieu 1979 apud Serpa 2013).

Mobility, as a form of communication and/or movement with social meaning (Cresswell 2006), is both a determinant and a consequence of Lefebvre's social practices that take place in the built environment. Moreover, mobility shapes the representations of space to a good extent: transportation axes structure not only the city's morphology but also people's fluxes, accesses, and actual possibilities (porosity), therefore, shaping signifiers and mental images. Finally, the spaces of representation relate to mobility to the extent that the imaginary about the city revolves strongly around its axes, the open spaces connected to them, and movement itself (Cullen 1961), as we can picture in and recall from Benjamin's flaneur, Cartier-Bresson's city photographs, The Beatles' Abbey Road and Penny Lane, and the hectic streets of Los Angeles in Ridley Scott's Blade Runner, to cite a few.

Thus, mobility and its absence are important parts of people's imagination and become more than cognitive coordinates; they become signifiers of action in the real world, informing common *citadins* as well as decision-makers. Mobility and immobility 'escape the bonds of individual dreams and aspirations and become social. They become political' (Cresswell 2006, p. 21).

10.4 Mobility, Porosity, and the Right to the City

In Lefebvre, 'the right to the city is designed to further the interests "of the whole society and firstly of all those who inhabit"' (Lefebvre et al. 1996, p. 158 apud Purcell 2002, p. 102). It sustains the very struggle of all territorially oppressed groups to free themselves and build another territory of possibilities through participation, starting from a new logic in the distribution of investment in housing, services, and mobility. Drawing primarily on Lefebvre's works, Purcell (2002, p. 103) argues that

> the right to participation rejects the Westphalian notion that all political loyalties must be hierarchically subordinate to one's nation-state membership [...] It proposes a political identity (inhabitance) that is both independent of and prior to nationality with respect to the decisions that produce urban space.

In other words, immigrants and 'marginal' populations share with elites the right to privileged city centres, 'instead of being dispersed and stuck into ghettos' (translated from Lefebvre et al. 1996 apud Purcell 2002, p. 102).

While in Lefebvre's geography of socially structured spaces, the first aspect of the right to the city concerns direct participation in planning decisions that accrue on city space, the second is the right to appropriation, including the right to physically access, occupy, and use urban spaces, i.e., porosity. In this sense, appropriation demands that collective and public policies consider the inequality of risks and opportunities that are historically, institutionally, and continually established in the urban landscape (Kobayashi and Ray 2000, apud Cresswell 2006, p. 165) and are positively engaged in a radical change that brings the disenfranchised into the decision-making spheres. Here, as Fraser (2001) reminds us, there must be a distribution of material resources that guarantees participants' independence, and institutionalized standards of cultural value that express equal respect for all participants and guarantee equal opportunities to achieve social consideration.

10.5 Towards a More Mobile, Accessible, and Porous City for All: Operationalizing Complementary and Somewhat Overlapping Concepts

Kaufmann (2014) emphatically proposes the use of 'concepts, methods and methodologies that allow us to equate [society's] outlines and problems' based on mobility as 'a whole social phenomenon' (p. 57); he then takes mobility to be the encounter of motility, connectivity, and reversibility, where motility would be 'the set of characteristics specific to an actor that renders him or her mobile [and] therefore refers to the social conditions of access, [...] to knowledge and skills, [...] and to mobility projects'. More specifically, it is 'the manner in which a person or group appropriates the field of the possible in terms of trips and makes use of it; it highlights therefore intentionality and self-determined projects' (pp. 61–62). Connectivity 'is spatially constituted in the near and the distant, or more precisely by the place of residence of those who use technologies and information and by the places that are brought closer through the speed of transportation. It differs according to one's relationships with space [and technology]' (p. 82), while reversibility, 'equally concerns travel times themselves, which are more and more used as whole social times in which multiple activities of leisure and work are undertaken. It also differs according to one's relationships with space [and technology]' (p. 84). While connectivity and reversibility are important variables, they are partly determined by technological, de-territorialized devices, as groups are more or less fond of them, and will not be the focus of this study.

Numerous definitions of accessibility depart from Hansen's (1959) classical definition, based on the 'potential [...] for interaction' (p. ii), while many others see it as a basic condition to reach 'opportunities available to each and every person' (Pucci and Vecchio 2018, p. 108). It has, however, been mostly used to measure the impacts of land-use and transport systems upon city inhabitants (Geurs and Wee 2004), underestimating the society–space dialectical interaction as a determinant of actual participation in and fruition of opportunities. Accessibility to territories may not be enough, as opportunities are not equally available to all those who physically reach them (Piazzoni 2020). In order to understand the difference between potential participation and actual fruition of opportunities, we need to understand the porosity of places that renders them more or less welcoming to different groups.

Secchi conceptualizes porosity first according to the proportion and distribution of open spaces environmentally qualified, available, and accessible in the city (including by public transportation) to varied flows of people, activities, and events of different origins. Second, he sees porosity as urban territories' potential for welcoming actors' projects (Secchi 2013), i.e., the hospitality they encounter in space, signalling that the number and physical accessibility to open spaces are not enough. The concept also amasses a phenomenological aspect that can only be understood when we take the *citadin* to be the actor, instead of the subject. In Stavrides (2018, p. 32), urban porosity is

> the sociotemporal form that an emancipating urban culture may take in the process of inhabitants reclaiming the city; [it] is activated by urban struggles and can become a form of experience that activates relationality rather than separation.

Summarizing, while motility—one of our specific concerns in this study—is a quality of the actor and/or of the dialectical relation between the self and the field of the possible, accessibility regards the concrete conditions given to people to potentially participate in activities depending on their motility, and porosity is a quality of the territory and/or of the dialectical relation between space and society that brings the *citadin* to actual fruition of activities. The three of them together and inseparably carry city dwellers' concrete and realizable possibilities concerning their projects in the territory.

One way to start picturing how the society–space dialectic based on motility, accessibility, and porosity shapes daily social relations, cognitions, and expectations is to listen to inhabitants, particularly the disenfranchised, in places where spatial justice is an issue.

10.6 San Siro (Milan) *Citadins*: How They See and Live the City

Most European metropolises have experienced for many years difficulties in dealing with high proportions of migrants in marginal, unequal neighbourhoods and with their movements across town (see Bourdieu 1993): they represent the threats that come with the foreign (Cresswell 2006), cultural clashes, and competition on the job market. These migrants usually dwell in places characterized by low social, symbolic, cultural, and spatial capital, where other outcasts have been living; together, they can be seen as the dominated in Bourdieu's force field. San Siro and its inhabitants seem to truthfully represent this dynamic.

Having been in the neighbourhood since 2013, 'Mapping San Siro' well summarizes its main features:

> One of the largest public housing districts in the city of Milan, [San Siro] is made up of 6135 dwellings and a population of approximately 12,000 inhabitants. [While it] is part of the

> consolidated city and is located in a rather central, comfortable and accessible context, [it can] be considered an 'internal periphery', by virtue of its exclusion from the development dynamics of the surrounding city [and] by the presence of numerous inhabitants of foreign origin (in total about 50% of its inhabitants) [and] of fragile populations, such as the elderly, often alone, and people with serious psychiatric pathologies [amid] numerous active and diversified local subjects, […] many empty spaces and numerous public spaces [that] represent an important heritage not only for the neighbourhood, but for the entire city(Mapping San Siro. *Il quartiere*. Retrieved Feb 18 2021 from (http://www.mappingsansiro. polimi.it/il quartiere).

The neighbourhood population is 51.3% Italian and 48.7% foreign. Crossing four data sources (Cognetti and Padovani 2018; Blangiardo 2009; Mapping San Siro 2019; Tuttitalia 2016), we can state that among Italians, the male population is 47.8% and women account for 52.2%, whereas among foreigners, these rates vary considerably: while among Egyptians and Moroccans, 68% are men, among Peruvians 66% are women. The working age population (here considered to be those between 25 and 64 years of age) is also fairly different across origins, pointing to the fact that while most migrants come to Milan in order to work (due to their age cohorts), a number of Italians who come to live in San Siro are off the job market (mostly the elderly). This can be translated into people having low-paid jobs and living on pensions, and means that little income comes into the neighbourhood.

Based on the available data and on gender, origin, and age cohorts, we designed a stratified sample of 100 inhabitants aged 15 or over (Table 10.1) in order to survey their mobility needs and conditions as well as their feelings towards Milan's porosity vis-à-vis their projects and personal profiles. The survey instrument was available in Italian, English, Arabic, and Spanish. The confidence-based and co-dependable social relations constructed through Mapping San Siro enabled the recruitment of two local and six foreign inhabitants that were trained to collaborate in interviewing the different nationalities according to their backgrounds. Their involvement was not only instrumental in approaching their peers, but also in translating meanings and cultural subjectivities back and forth, enhancing an already open dialogue. I, as the main researcher, interviewed forty-five people belonging to different cohorts. Interviews were conducted between October 2019 and January 2020.

The survey questions ranged from reasons for frequency and destination of usual trips outside San Siro, usual choice of and opinions on public transportation, the importance of moving and personal abilities to move around town, personal agency and projects, to perceptions of being welcome and able to participate in activities in the city (porosity proxies), as well as questions related to interviewees' profiles. The main results concerning how motility, accessibility, and porosity shape daily social relations, cognitions, and expectations show that 62% of interviewees travel outside the neighbourhood seven days a week and 65% have work/study as their first motive for this, while only 10% declared that they travelled for culture/leisure/sports and nature purposes. Of those travelling, 82% percent use public transportation and 13% find social relations to be the worst problem during their trips (e.g., religious intolerance, xenophobia). Moving around town is totally important or important for

Table 10.1 Survey sample of San Siro inhabitants by age, gender, and origin

Age	15–24	25–44	45–64	65+
Italians				
Men	3	8	8	6
Women	2	8	9	8
Filipinos				
Men	1	2	1	
Women		1		
Egyptians + Moroccans				
Men	2	12	2	
Women	1	6	1	
Peruvians				
Men		1		
Women		1	1	
Others				
Men	1	6	2	
Women	1	6		

Source The author

81% due to factors such as daily necessity, psychological and physical well-being, and meeting people, as Table 10.2 shows. The data seem to indicate a habitus forged by necessity, typical of the working classes, including few interests/opportunities to participate in cultural and leisure activities. While 31% are native to Milan, 42% did not decide to live in Milan on their own and had somebody else deciding it, showing little agency or empowerment in the face of hardship.

The neighbourhood, only five kilometres away from the Piazza del Duomo, is well served by public transportation, with two metro, one tram, and numerous bus

Table 10.2 Why moving around town is important

	Frequency	Valid %	Cumulative %
Daily necessity	40	40.0	40.0
Psychological well-being	18	18.0	58.0
Meeting people	7	7.0	65.0
Physical well-being	7	7.0	72.0
Little important	7	7.0	79.0
Does not know/no answer	9	9.0	88.0
Not important at all	12	12.0	100.0
Total	100	100.0	

Source The author

lines within walking distance; 82% of its inhabitants use these services on a regular basis, 56% take up to 30 min to reach their most regular destination, and 58% walk less than ten minutes in these trajectories. Accessibility, thus, does not seem to be a problem.

Only 16% have moving restrictions, and 15% due to physical difficulties. Other difficulties are: 17% admit they do not speak Italian well, 10% have a hard time understanding maps, and 66% do not regularly use apps for moving purposes, revealing motility issues due to personal characteristics. Only 51% have or had one or more personal professional projects or dreams when young or when they moved to Milan, out of which 85% think that obstacles to moving around town may have hindered their projects or dreams 'in very few or no opportunities'. When asked if moving difficulties hindered other aspects of their lives, 84% of all interviewees answered 'in very few or no opportunities'. While this may indicate a highly accessible city, where the elderly and migrants can reach and potentially participate in activities, it actually seems to reveal a habitus based on very low expectations as a result of few embedded opportunities. When asked about the concerning obstacles, a few disturbing answers came out: language (2 respondents), fear (1), presence of 'Arabs' (1), and feeling like an outsider (1). While 79% feel welcome enough outside their neighbourhood, 21% feel little/very little/not welcome at all; among them, only six out of 100 interviewees declared that it hampers their lives to a sizeable extent or totally. Thus, if porosity is a problem, respondents did not voice it.

A selected predominant profile of users, which pictures the mode (most common answers), is shown in Table 10.3.

The variables and predominant profile that emerge from this quantitative effort also need to be contextualized by the qualitative effort, given researchers' participant

Table 10.3 Predominant profile

Age 25 to 44 (1)	42%
Male (1)	53%
Non-qualified profession (2)	17%
Travel 7 days per week	63%
Use public transportation	82%
Work is the main reason for travelling	58%
Moving around considered totally important	46%
Speak Italian	82%
Born in Milan	31%
Had no professional projects/dreams	43%
No or few mobility difficulties that hinder projects	85%
No or few mobility difficulties that hamper life	84%
Feel totally welcome outside the neighbourhood	55%
Not feeling welcome does not hamper life at all	36%

Notes (1) Statistically pre-determined. (2) According to the Italian Ministry of Labour categories observations and the literature. The San Siro public housing area is fairly deprived of good urban structures such as qualified open spaces and public or private cultural, sports, leisure, and commercial/service facilities, configuring a different social space in Central Milan. It is characterized by low spatial capital and treated like a ghetto in the local press, as Purcell (2002) would note. Its inhabitants may not be very proud of it, as many say moving there was not a personal choice, youngsters want to spend leisure time outside its limits, caring for public spaces has been a hard-to-build attitude, and most would not let themselves be photographed in them. Participants in neighbourhood meetings say they feel ignored as far as their rights as taxpayers go, while they feel under surveillance as outcasts, in a true system of differences.

A quali-quantitative understanding points to respondents who are working people in low qualification jobs, whose habitus is based on surviving strategies, rather than on professional dreams and consumption freedom. When asked who decided to move to Milan, a respondent answered: 'poverty'. Their habitus is also a result of low expectations: another interviewee stated that 'poor people don't have professional dreams, they just want a job'. They enjoy a very good level of accessibility, but not motility, in face of the little room there is to want/wish for anything within their habitus and their intrinsic social and cultural difficulties, especially for those with little formal education and for Arab women. I interviewed an Eastern European woman in her forties who could not read or write and I could not reach an Arab woman to conduct the interview because she could not go out or have contact with strangers, said her teenage daughter.

Not enjoying full citizenship (Purcell 2002), foreigners are experiencing prejudice due to origin and/or looks—a darker skin, a veil, a different dress code, a diverse eating habit, as put by a 65+ Italian lady: 'I want nothing to do with foreigners; they are the ones who want integration, not me'—or due to simple or disarranged clothing caused by poverty and/or mental illness. Two consequences from not being recognized (in Fraser's 2001 sense) seem reasonable: (1) the city's porosity is low to them, as they are not cherished as valuable citizens in high-profile public spaces and in most consumption-geared places, as shown by Piazzoni (2020). 'I am not very welcome, but I travel in order to see my friends; so, it doesn't hurt', says one Sri Lankan young man; and (2) foreigners develop position homology (Bourdieu 1996) or blatant denial, either following Italian styles and behaviour and/or talking about themselves as if they were ethnically Italian: 'I used to face a lot of rejection, people stared at me... What changed? I changed my looks. It never happened again', says a successful Eastern European woman in her fifties.

Perhaps, maybe as a strategy, declaring to feel welcome and not being hurt by prejudice helps them endure difficulties and envision a different future. That, however, weakens their sub-dominant cultural bonds (Serpa 2013) and the articulation of under-commons spaces and ties (Sheller 2018a)—like the Bangladeshi do in Rome (Piazzoni 2020)—and hinders constituent processes of commonality (Hardt and Negri 2014) that could lead to a city of rights for all. On the other hand, a number of interviewees voice their uneasiness with the city: an established Syrian, with whom I became better acquainted, answered all questions herein discussed very positively; when the questionnaire was over, he started telling me how unwelcome he feels in Milan, a city that is very inaccessible to foreigners—very different from Naples, where his family has Italian friends and goes for vacation. Likewise, an Italian middle-aged man with a light mental illness, who, when asked how he liked working at the Duomo Plaza, said: 'Humpf... They pay me'. Others feel uneasy talking about their hardships, as reported by an Arab interviewer: 'They don't want to talk. It is as if they were paying you [to get the interview over with]'.

Regarding the 65+ population, the survey sample indicated a total of fourteen Italians, of which eight women. Eleven declared to use public transportation and six travelled on a daily basis; nine think it is important or very important to move around town, mostly to meet other people or to seek public services they can reach within 30 min. However, six (43%) face health restrictions to mobility. Only seven (50%) used to have reasonably paid jobs. Although 65% did not decide to move to Milan on their own, they all feel welcome across town, and only two elderly thought mobility had hindered "somewhat or a lot" their juvenile professional projects. In other words, this is a low-paid, low-expectations, sort of satisfied group. Or this is what they voice, feeling too old to face failures or wish otherwise, as indicated by a once well-recognized artist who loves to be invited out but felt 'trashed' by society as he grew poor.

With the pandemic under control, the research was resumed in 2021 and its results are expected soon. In any case, a number of issues seem to have been continuously confronted through the Mapping San Siro Lab and its partners since 2014, with works towards diagnosing and designing public policies for housing, jobs, and culture valorisation—which enhance inhabitants' pride and self-esteem—and efforts to empower foreign women in order to engage them in community life, among many other things. As a bottom-up, underfinanced effort, its actions are very important but still limited to a few, and require time to be recognized by the majority of inhabitants.

The picture presented here is a snapshot, a working scenario which helps local initiatives—particularly neighbourhood-based projects in marginal areas—understand and focus on the most problematic, sometimes underlying aspects of marginality. Understanding motility as the main determinant of urban mobility together with accessibility and porosity leads us to question whether the marginalized can, beyond moving with comfort, safety, and efficiency, enjoy destinations in their fullness and according to their potential for sociability, education, and cultural recognition and insertion 'in their own terms' (Miraftab 2009). In a context of little space capital (Soja 2010), low self-esteem, and ethnic stigmatization, low expectations and self-containment can be expected, all concurring to a low motility/low porosity picture, in which social injustice hits harder. While our quantitative results seem to indicate satisfactory results concerning San Siro inhabitants' mobility, a number of qualitative remarks show that problems can be deeper, not yet surfaced or voiced.

If Sheller (2018b) is correct in indicating that the marginalized are key to the process of establishing mobility (and thus social) justice, as 'strangers [who] invent new ways of communicating with others, new ways of acting together, new places of meeting and being together' (p. 161), and if Stavrides (2018) is also correct in saying that urban struggles are at the basis of a porosity that will emancipate the marginalized, we need to act on the socio-geographical inequities and with the stigmatized immediately. Hail the bottom-up initiatives!

References

Barbosa JL, Dias CG (2013) Solos culturais. Observatório de Favelas

Blangiardo, Gian Carlo (2009) L'immigrazione Straniera in Lombardia. Rapporto 2008. Iniziative e Studi sulla Multietnicità. Osservatorio regionale per l' integrazione e multietnicità, Lombardia

Bourdieu P (1979) La distinction: critique sociale du jugement. Paris: Minuit

Bourdieu P (1993) La misère du monde. La Seuil

Bourdieu P (1996) Razões práticas: Sobre a teoria da ação. Papirus

Bourdieu P (2000) Pascalian meditations. Stanford University Press

Cognetti F, Padovani L (2018) Perché (ancora) i quartieri pubblicci. Un laboratorio di politiche per la casa. Milano: Franco Angeli

Cresswell T (2006) On the move: mobility in the modern Western world. Routledge

Cullen G (1961) Townscape. Architectural Press

Davis M (1993) Cidade de quartzo: Escavando o futuro em Los Angeles. Página Aberta

Fernández-Álvarez R (2017) Inequitable distribution of green public space in the Mexico City: an environmental injustice case. Economía, Sociedad y Territorio 17(54):399–428

Fraser N (2001) Da redistribuição ao reconhecimento? Dilemas da justiça da era pós-socialista. In: Souza J (ed) Democracia hoje. Editora Universidade de Brasília, pp 231–239

Geurs KT, van Wee B (2004) Accessibility evaluation of land-use and transport strategies: review and research directions. J Transport Geogr 12:127–140

Hansen WG (1959) How Accessibility shapes land use. J Amer Inst Planners 25(2):73–76

Hardt M, Negri A (2014) Declaração—isto não é um manifesto. Nn edições

Harvey D (2004) The 'new' imperialism: accumulation by dispossession. Socialist Register 40:63–87

Kaufmann V (2014) Retour sur la ville. Presses Polytechniques et Universitaire Romandes

Kobayashi A, Ray B (2000) Civil risk and landscapes of marginality in Canada: a pluralist approach to social justice. Can Geogr 44:401–417

Lefebvre H, Kofman E, Lebas E (1996) Writings on cities. Blackwell

Mapping San Siro (2019) Istantanee di San Siro. Presente e futuro del quartiere. Retrieved Feb 17 2020 from www.mappingsansiro.polimi.it/wp-content/uploads/2019/02/

Miraftab F (2009) Insurgent planning: situating radical planning in the Global South. Plan Theory 8(1):32–50

Piazzoni F (2020) Visibility as justice: Immigrant street vendors and the right to difference in Rome. J Plan Educ Res 1:1–16

Pucci P, Vecchio G (2018) Mobilità e inclusione sociale. Pianificare per vite sempre più mobile. In Atti della XXI Conferenza Nazional SIU: Confini, movimenti, luoghi. Politiche e proggeti per città e territori in transizione, pp 107–116

Purcell M (2002) Excavating Lefebvre: the right to the city and its urban politics of the inhabitant. GeoJournal 58:99–108

Santos M (1996) A natureza do espaço: Técnica e tempo, razão e emoção. Hucitec

Secchi B (2011) La nuova questione urbana: Ambiente, mobilità e disuguaglianze sociali. CRIOS 1:83–92

Secchi B (2013) La città dei ricchi e la città dei poveri. Laterza

Serpa A (2013) O espaço público na cidade contemporânea. Contexto

Sheller M (2018a) Theorising mobility justice. Tempo Social, 30, 2, pp 17–34

Sheller M (2018b) Mobility justice: the politics of movement in an age of extremes. Verso

Soja E (2010) Seeking spatial justice. Minnesota University Press

Stavrides S (2018) Urban porosity and the right to a shared city. In: Wolfrum S (ed) Porous city: from metaphor to urban agenda. Birkhäuser, pp 32–37
Tângari V (2018) Open Space Systems in Rio de Janeiro: The public and private spheres reflected in the urban landscape. In: Capanema-Alvares L, Barbosa JL (eds) Urban public spaces: From planned policies to everyday politics (illustrated with Brazilian case studies). Springer, pp 109–126
Tuttitalia (2016) Statistiche Demografiche—Città Metr. di Milano. Popolazione per età, sesso e stato civile 2016. Retrieved Oct 07 2019 from https://ww.tuttitalia.it/lombardia/18-milano/statistiche/popolazione-eta-sesso-stato-civile-2016

Chapter 11
Peterbos: Living in the Park, Inhabiting the City

Paola Viganò, Bertrand Plewinski, Guillaume Vanneste, and Nicolas Willemet

Abstract This paper explores the urban issues underlying the design experience in the Peterbos neighbourhood, Anderlecht, Brussels-Capital Region. It presents four themes, based on the living experiment of this urban project, which consists of a master plan for the renovation of public spaces (Studio Paola Viganò and vvv architecture urbanisme 2020). It starts with the critical perspective of a 'project for the ground'. As an embodiment of modernity, collective living in high-rise buildings has made it possible to free up a large area of ground for use as a shared landscape. In Peterbos, this large ground has aged, deteriorated, and become disconnected from the city. Up until now, these characteristics have made Peterbos a place where all the 'misery in the world' (Bourdieu, La misère du Monde. Seuil, 1993) has been concentrated. A long transformation process is now underway: the renovation of housing and public spaces proposes new living conditions and a new image for the district. However, there are still questions about the appropriateness of such an investment in the absence of a radical rediscussing of what makes Peterbos an enclave for the poorest. Our analysis starts with the ground of Peterbos and its relationship with water flows, biodiversity, and the rest of the city. The modern project focuses on the liberation of the public ground. We see the Peterbos project as an opportunity for critical reinterpretation. Second, we reconsider the district's position in the city and the need to reverse feelings inside and outside, aiming to renew relations with the metropolis. Third, a broader understanding of the environment is necessary in order to take part in an ecological transition. The notion of diversity and mixed-use as a fertile

P. Viganò (✉)
EPFL, IUAV Venice, Studio Paola Viganò, Lausanne, Brussels, Belgium
e-mail: paola.vigano@epfl.ch

G. Vanneste
Studio Paola Viganò, Brussels, Belgium
e-mail: guillaume.vanneste@uclouvain.be

B. Plewinski
UCLouvain-LOCI, VVV Architecture Urbanisme, Brussels, Belgium
e-mail: studiobrussels@studiopaolavigano.eu

N. Willemet
ULiege, VVV Architecture Urbanisme, Brussels, Belgium
e-mail: info@vvvarchitectes.be

N. Aernouts et al. (eds.), *Urban Living Lab for Local Regeneration*,
The Urban Book Series, https://doi.org/10.1007/978-3-031-19748-2_11

framework for emancipation and individual initiative is then discussed. Finally, in the conclusions, having explored the *progetto di suolo* as a manifold agent and pushed it to its limits, we conclude by examining the 'stone guest'. Indeed, urbanism and investment in urban renewal do not represent an autonomous and self-responsible solution to the social and urban challenges society is currently facing. The design of public space represents a wide, but also a narrow, space for manoeuvre. When structural changes are implied, they do not tackle the basis of inequality concentrated at this site, linked to decisions made in the past that do not show the expected results. Interaction with economic policies is still too weak. We choose to use clear but sometimes burdensome vocabulary to discuss these spatial and social matters, not escaping the difficulty of the topic. All the same, urban and landscape designers have a responsibility and the possibility to assert the original meaning of politics as the organization of public life in the city, more broadly addressing actions in space and measures to reduce inequality and restore the dignity of the people who live there.

Keywords Brussels · Peterbos · Urban design · Urbanism · Ground · *Progetto del suolo* · Social housing · Housing policy

11.1 Introduction

Peterbos is a park in which people live. It is a modernistic neighbourhood built in the second half of the twentieth century in Anderlecht, Brussels. As an embodiment of modernity, collective living in high-rise buildings freed up a large area of public ground that was envisaged as a shared landscape where community relations could be enhanced. In Peterbos, this large surface area has aged and deteriorated. It has lost its services and facilities and is now socially and spatially disconnected from the city. A long transformation process is currently underway: the renovation of housing and public spaces aims to finally give the district a new image and quality.

This article describes the first steps of the ongoing project and research, developed by two architecture and urban design firms—Studio Paola Viganò and vvv architecture urbanisme (2020)—within the framework of a Sustainable Neighbourhood Contract. This is an urban renewal policy based on a regional subvention process—from the Brussels Region to the municipality, in this case Anderlecht—which is applied to a four-year programme of urban and socio-economic projects within a specific local perimeter in Brussels (Degros and Decleene 2014). In this case, the perimeter of the Sustainable Neighbourhood Contract literally corresponds to the so-called 'Cité du Peterbos', officially named 'Parc du Peterbos'.

11.2 The Misery of the World/Space for Manoeuvre

Peterbos has the lowest labour force participation rates in the region and the highest rate of unemployment, especially among young people, which currently makes it an area of extreme poverty. Until now, these characteristics have made Peterbos a place where, as Pierre Bourdieu pointed out in the 1990s, 'the world's misery' has become concentrated. Inevitably, in this context the question arises of what kind of project might make sense. A philanthropic goodwill project? A real political project carried out by all players around the table? As architects and urbanists, our focus and our only territory of action is, in this case, public space. Bringing together matters of architecture, landscape, urban design, and sociology, the project deals with the public ground as a prior object which is then considered through various understandings, tools, and actions. Our design starts with the ground (Fig. 11.1): with its relationship with water flows and biodiversity, the ground floors of buildings, and the rest of the city, and its ability to support programmatic diversity. It is a space for manoeuvres on both a wide and a narrow scale. Wide, because it fulfils a plurality of practices, ordinary and extraordinary: public space concerns everyone. Narrow, because our project does not have the power to transform the inequality concentrated at this site, which is linked to past decisions that have turned out to be ineffective from the perspective of social integration. Interaction with economic policies is still too weak.

Urban design and investment in the refurbishment of the spatial environment is not an autonomous and self-responsible solution to the social and urban challenges our society and poor neighbourhoods are facing today. Design claims that the core meaning of politics is the organization of public life in the city to more broadly implement actions and measures to reduce inequality and restore the dignity of the people living there. In such a context, public space can fully express its potential as an emancipation device, support, and capital on which to build.

11.3 Disclosing the Ground

To further elaborate on the ongoing project, we will conceptually break down the ground of Peterbos into four different themes, adding layers of complexity in order to start envisioning the multifactorial meaning and possibilities of this entity. The first section deals with the modern heritage of the ground and its possible reinterpretation. The second chapter develops the 'project of ground' (Secchi 1986) strategies and their potential in relation to the third part, which establishes the ambition to aim for a societal and ecological transition. The fourth chapter considers public space as a representative space of emancipation, respect, and diversity (Fig. 11.2).

By disclosing the ground project strategies, we seek to answer this question: how can Peterbos' ground be renovated to support its residents in becoming a fully integrated part of a larger urban context, while reinforcing the significance of the park as an exceptional qualitative context in a rapidly growing metropolis and establishing

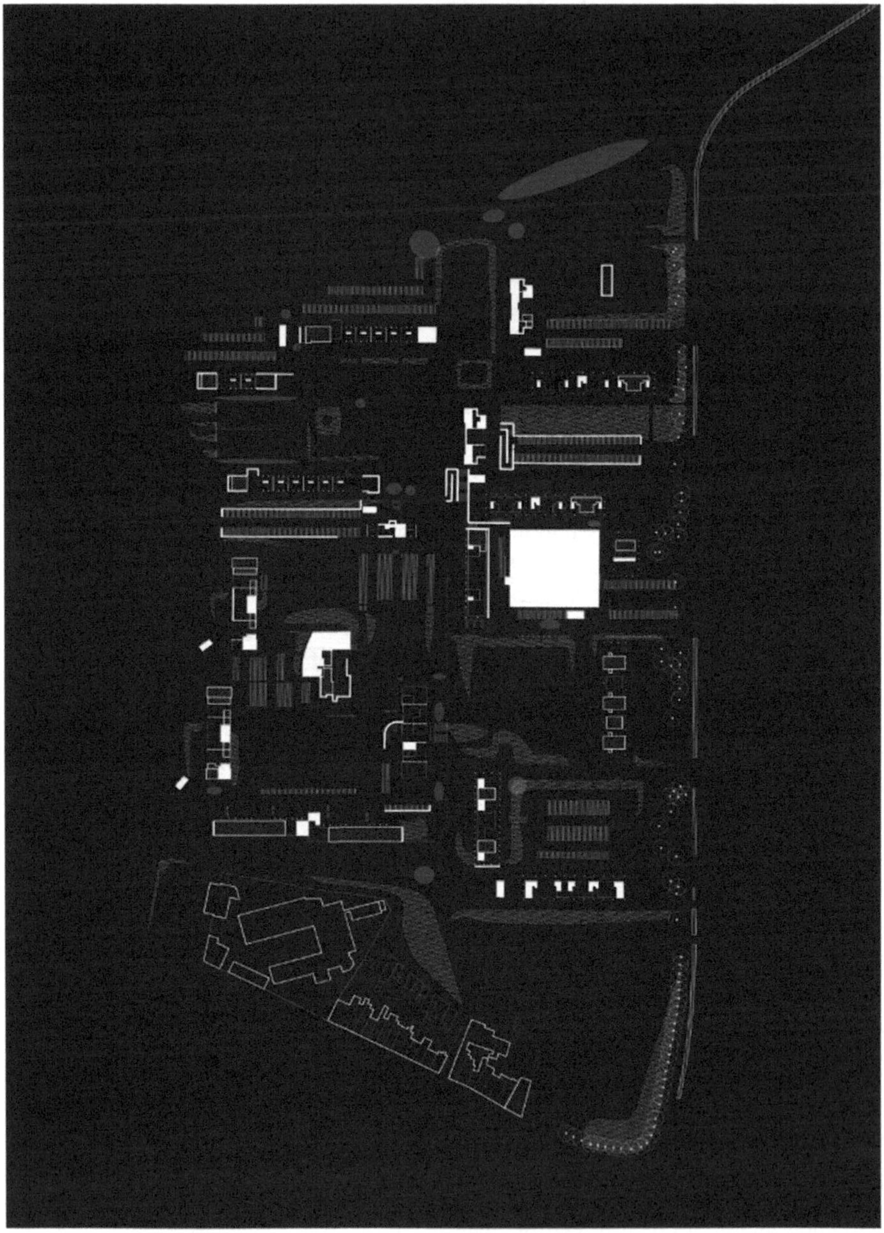

Fig. 11.1 The existing ground as an ensemble of barriers, obstacles, and topography

Fig. 11.2 Images from the master plan

permeable borders with the rest of this metropolis in order to experience the public space of Peterbos? How can the design and renovation of Peterbos' ground transform it into a living and inclusive part of the metropolis? What is it unable to solve that should instead be tackled by structural changes in socio-economic political programmes?

Our conclusion revolves around three main understandings: (1) the recognition of the richness of the original project, highlighting its potential qualities (a ground monument); (2) the recognition that a renewed alliance between spatial design and economic development theories (and between spatial design and enlarged public investments and urban renewal policies) is necessary; (3) the strength of public space when it is intended not only as a physical feature but also as a tool to reformulate larger societal issues, and its design as a device to deal with the stone guest.

11.4 Reinterpreting the Modern Ground

11.4.1 Brussels' Transformation in the 1960s

The 1,600 apartments of the eighteen housing blocks of Peterbos were built between 1968 and 1980 under the impetus to modernize Brussels. Iconic operations—such as the universal exhibition 'Expo 58'—and massive infrastructural investments, such as the north–south railway connection (1952) or the development of urban highways in the core of cities, mostly represent the iconic embodiment of a modern Brussels.

The housing question is less valorized or less able to sustain a powerful narrative about a productive modernity in a still conservative context but was nevertheless largely part of the modern transformation of Brussels. Contrary to other European countries and for reasons we cannot examine here, modern housing experiments were both somewhat delayed (the first garden cities were built in the interbellum) and took place in a less hierarchic centre/periphery relationship, with the formation of a specific form of *banlieue radieuse* (Leloutre 2020; Smets 1986).

These districts are made up of housing blocks based on industrial and standardized designs, with large green spaces as common areas. Developed by real estate or social housing companies on cheap agricultural land on the city's outskirts, the authorities were in charge of building the roads, public infrastructure, and facilities. Mostly unfinished, these developments did not have sufficient community facilities or production equipment to become full-fledged neighbourhoods and instead became monofunctional enclaves (Sterken 2013). Both the city centre and the 'second ring' around Brussels were the locus of large housing projects during postwar urban development. Although some of them were conceived with careful attention to new ways of living with high spatial and landscape qualities, they still experience a variety of difficulties regarding their capacity to attract urban quality or embody a housing ideal and are therefore rejected from the collective imagination.

Peterbos is one of the peripheral developments of Brussels' 'second ring'. It was a new neighbourhood in Anderlecht which, according to the municipality's communication, constituted an 'open order, with social and commercial facilities on the ground floor, roads limited to the strict service of the residences and underground garages, and above all, a large proportion of green spaces extending further into a green park system that our municipality strives to create across its territory' (Commune d'Anderlecht 1963).

11.4.2 A Long Time Span

The first phase of the construction of the Peterbos site (blocks 1 to 8) was carried out according to the project of the modernist and functionalist architectural firm 'Groupe Structures', founded by the architects Louis Van Hove, Raymond Steniers, Jacques Boseret-Mali, and Jacques Vandermeeren. Groupe Structures is well known in Brussels for social housing projects similar in typology and landscape, such as Rempart des Moines (1964–1966), Potiers I and II (1958–1965), and the 'Cité Modèle' of Heysel in Laeken. For the latter, an architect of Groupe Structures collaborated with some of the best modernist Belgian architects, such as Renaet Braem, Victor Coolens, Groupe l'Equerre, René Panis, and Jean Van Dosselaere (1954–1955,1975) (Sterken 2013). The 'Cité Modèle' is composed of a set of towers and housing blocks grouped around a central high square surrounded by a pergola. It has car parks along the main roads, a monumental staircase connecting the existing city to the new city and its high square, and, finally, large green spaces and orthogonal paths between the towers. As for the ground, the high square is positioned on top of an underground car park, and the topography is partly made up of soil backfills. In the Rempart des Moines project in downtown Brussels, the blocks are laid out linearly in a north–south direction leaving a trapezoidal gap as public space. The ground here is the extended space of the street, made up of pavements which open up here and there with greenery or playgrounds (Strauven 1983). The approach is the same, however, as that of Peterbos: monumental ground for slabs and towers.

The Peterbos project was built over a long period of time. The first phase of the original design of Peterbos (blocks 1 to 8) has similarities with the other two projects by Groupe Structures. Car parks surround the housing blocks, accessible from the three roads and highways around the site (Boulevard G. de May, Rue de l'Agronome, and a former highway that is now a green walkway: *La promenade verte*). The ground at the foot of the towers is filled in and levelled to make a common plateau for blocks 1 to 7, at some points creating differences in height between the pedestrian area and the parking areas. The flat area of the central square is bordered by a canopy and a water basin (unbuilt but visible in the initial plans). The topography is shaped through architectural devices such as low walls, taluses, slopes, and monumental staircases. The high square is a plateau intended as common ground for residents and is situated at a distance from the city itself. The design therefore radically applies the functionalist principles of flow separation.

Despite the housing qualities in relation to the landscape, the initial intentions are lost in the southern part. The construction of the later phases appears to have been more fragmented and less attentive to the relationship with the ground, even though it had to deal with a steeper slope than the first phase of Peterbos, which is located on the flatter part of the site. The shared device of the common square of the first phase is replaced here by autonomous or twin blocks designed to accommodate the slope at their base (blocks 10 and 11, 17, and 18; framing tower 16). Tower 16, positioned on the steepest slope, has a massive base disconnected from the ground. Car parks were built mostly between the housing blocks, reducing space for pedestrians. Over the years, pedestrian space has been carved out of what was left by the car parks.

Reinterpreting the imaginary of the ground on a larger scale, the Peterbos landscape fits into the park system of Anderlecht, a collection of green and public spaces built both on the geographical morphology and on the green belt for the Brussels metropolis, set up at a communal level by Mayor Bracops (1947–1966) and implemented by municipal engineer Georges Messin (Leloutre 2017). In that sense, the park of Peterbos can be understood as an inhabited park within the Scheutveld park, established in the same period (1968). The landscape qualities remain throughout the site: the artificial 'plateau' to the north, now linked to the *balade verte*, a green walkway; the large spaces between the towers with various facilities (car park, football field, playground); and diversified vegetation. The juxtaposition of these landforms is the common ground of Peterbos. Over the course of the twentieth century, this base-ground embodied the different imaginaries that were attributed to them: the modern belvedere as a rampart against the old city in 1960s, and the car park floor in the late 1980s and 1990s. Today, other images can be evoked to describe this *monumental base-ground*, which reveal the mix of nonorthodox influences in the modern green carpet, inspired by Italian Renaissance garden villa projects (Fig. 11.3): the monumental staircases of the Boboli gardens (along Peterbos' central path); the green embankments that symmetrically frame the entrances to the Villa d'Este in Tivoli (entrance from the western boulevard in Peterbos); and the staircases as a backdrop to the Palazzo Farnese in Caprarola or the Villa Medici in Castello. This imaginary invokes another relationship with the ground: that of the garden, the monument, and the contemplative space from where one can extend one's gaze. This sophisticated approach is not surprising, although rarely or never evoked in the reading of Peterbos' urban and landscape structure. By reinterpreting these devices in the light of others in the history of architecture, a possible requalification strategy is initiated, aiming to transform physical or visual 'barriers' into potential places that reinforce the link with the natural ground, the slope of the valley.

11.5 Inside–Outside

Unlike most other European cities, Belgian cities and especially Brussels never implemented the project of a segregated periphery, de facto installing spatial injustice within its metropolitan space via a *banlieue* relegated to the 'outside' of the city, at a distance from the core. On the contrary, several social housing neighbourhoods can be found in different places in the city centre or within the 'first Couronne', the municipalities located directly around the centre. At a metropolitan scale, this sparse location of modern social housing units could be evaluated as a benefit for accessibility, social mixing, and spatial justice. At a local scale, there are still strong contrasts between these neighbourhoods and their direct vicinity. Even the rupture in urban fabric typologies, as well as the concentration of a poorer population, leads to a strong fragmentation of space, the feeling of a patchwork, and a barrier effect.

Giusto Utens
The Pitti Palace and Boboli Gardens, 1602

The central path and its monumental staircase

Jean-Baptiste-Camille Corot
View of Florence from the Boboli Gardens, 1835

View from a roof towards the city

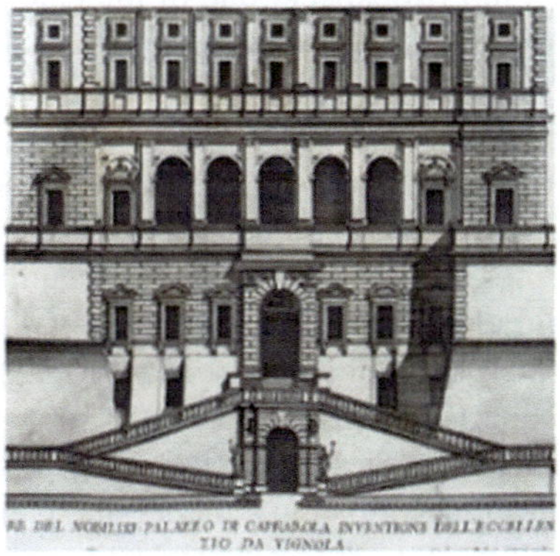

Giovanni Battista
Farnese Palace in Caprarola, 1655

Stair and ramp system from the boulevard

Fig. 11.3 Reinterpreting the modern ground

The modern urban language appears radically contrasted in Peterbos: the eighteen blocks and towers are surrounded by garden cities or postwar urban development. Although from an architectural point of view, important differences between successive construction phases of development inside Peterbos should be mentioned, the image reflected is that of a homogeneous, poor, and enclaved neighbourhood. The concentration of high-rise buildings and socio-economically poor populations, the lack of facilities, workplaces, jobs, and an 'urban feeling' create the figure of an outside area of the city, a peripheral space, right at the core of the metropolis. As an island, Peterbos has created strong and visible borders and gates that have to be crossed. As an island, one might also be inclined to bypass this area during daily movement in the city in order to avoid staying there or passing through it.

The project deconstructs this reversed figure of an outside inside the city, while maintaining the qualitative and even aesthetic specificities of the landscape and urban design of Peterbos. The challenge is to preserve what is beautiful about the island but at the same time increase the number of connections to open the site up to the city and vice versa. To deal with this morphological, topological, and social issue, the design of the open space took several approaches, rooted at different levels in the strong figure of ground continuity: describing the geography of the Broekbeek valley and valorizing the former monumental ground, while domesticizing it (1), extending the 'centre' to create a welcoming threshold between the site and the city (2), requalifying the *plaines* as public spaces (3), and proposing to name the streets and squares to weaken the stigma of living in Peterbos (4).

First, the recognition of the local geographic situation and the steep slope from the crest to the low wet point of the Broekbeek are initial elements to be addressed in most decisions about paths, flows, and spaces. Paths are designed to either find their way *en corniche* or go down the slope. Furthermore, the connection with the *balade verte* on the crest—a green route aimed at connecting all major green and open spaces on the outskirts/second Couronne of Brussels—and the Broekbeek, one of the catchment rivers and tributaries of the Senne valley that shapes the topography of Brussels and historically structures the old faubourg of Anderlecht, will make Peterbos part of a metropolitan system of parks. Second, the main entrance is reconceived as a primary link with the boulevards, a linear urban structure from the mid-twentieth century that crosses Anderlecht (the continuous string of Boulevards Paulsen, Herbette, and Graindor) (Fig. 11.4). Third, a system of *plaines*/squares (now partially encumbered by parking lots) adds diversity and different gradients of urbanity and, again, thresholds. The new equipment (developed via specific architectural competitions) will complete the improvement of the *plaines*/squares project. Fourth, the strategy of renaming the streets attempts to reinvent the notion of an address through workshops with residents. The idea is not to recreate the street following a model that is inadequate here, but rather reappropriate the place and not get lost. It seeks to reconnect the buildings to their ground.

These first steps aim to provide clear guidance which, as a whole, encourages the use of other forms of mobility than those planned in the 1960s. It links the transformation of space to the evolution of modern project imaginaries and, in parallel, to changes in mobility practices. More walking and less car usage can help reverse the

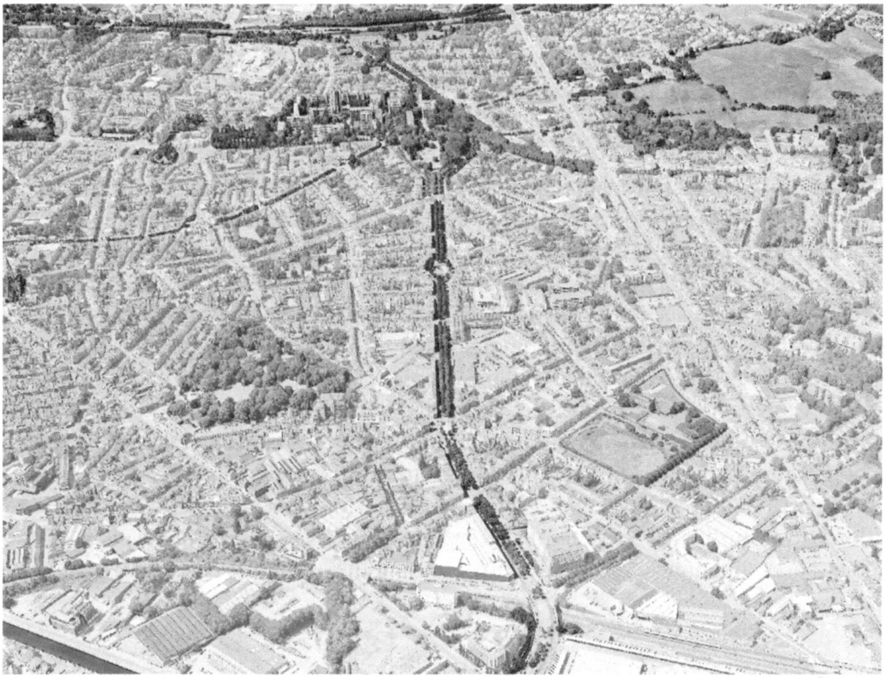

Fig. 11.4 Inside–outside

outside and inside image of Peterbos (currently with parking spaces as thresholds between the two). The ground is the locus of this reversal, addressing the open public space per se and in the city.

11.6 Setting the Scene for the Ecological Transition

Ninety per cent of the surface area of Peterbos is open, potentially green, and permeable. Here, soil and water are two transformative agents such that the design of public space deeply embodies in the ground. The renewal project represents an opportunity to propose an eco-systemic and metabolic shift for Peterbos. *What if* Peterbos was actually better equipped than traditional urban tissue to face climatic issues and climatic transition? *What if* Peterbos could behave as a 'climatic body'? Could the large open ground play a key role in supporting this ecological and urban transition?

11.6.1 Resizing Car Parks and De-Sealing

A fundamental evolution is needed with regard to the 'space of the car'. Today, Peterbos appears to be a gigantic car park, with the juxtaposition of an asphalt carpet at the foot of the buildings. The project reverses the dominant status of cars and the car-oriented design of the surrounding area by reorganizing the erratic spatial organization of the parking lots and considering reduced future car use in Brussels, included in the Good Move regional policy for mobility (Bruxelles Mobilité 2020). This reduction in car usage and parking spaces will go hand in hand with better overall pedestrian access to the site, enhanced connections to public transport, and direct paths between car parks and housing entrances. However, the proposed design maintains accessibility by car and a number of parking areas close to each building, as strongly advocated for by residents, including some women, who drew our attention to feelings of insecurity when moving around within their own neighbourhood. This is a preliminary step to allow other environmental strategies to take shape on-site, since part of the parking areas will be assigned to other purposes. This reduction allows for a further de-sealing of the ground, literally reducing the asphalt surface of car parks and unnecessary car roads crossing the site.

11.6.2 A Climatic Body

Increased permeability would then promote the idea of Peterbos as a climatic body, regulating atmospheric and flow parameters. Climate change presents us with extreme situations that an open space like this can help us tackle: increasing the cool island effect at an even larger scale (easier here due to the large vegetation patches than in the dense city), allowing water infiltration, and increasing ecological continuity. For instance, the design proposes diverting the rainwater drainage system from impermeable surfaces of the sewage system and guiding it into ditches and swales where the water has the space and time to infiltrate, reducing the downstream flood risk. This image of the ground as a climatic body could assist with the envisioning and reconnecting of some previously disconnected elements, for instance helping the built elements and their users be part of this system. As a carrier of flows, matter, species, and the atmosphere, the ground can offer a transformative landscape of wet and dry areas, slopes and plains, maintained areas, and others left to lie fallow.

11.6.3 Maintaining the Landscape of Ecological Transition

From a reductionist modernistic vision of the ground as an aesthetic environment to a relational reappropriation of the resources, along with the 'wild' uncultivated landscape comes the reclamation of the ground as a resource. The heritage of community gardens and urban struggles shows that there is a long history of family gardens as collective and social objects, which are constantly at risk and have a dynamic geography (Cahn et al. 2018). The master plan proposes cultivating the ground of Peterbos in some areas, creating new value: as support for not only vegetable and food production, but also and mainly as a social catalyzer. Small structures will be developed in collaboration with local associations, such as a greenhouse and apiaries. Furthermore, this is an opportunity for new synergies with the ground floors of the buildings, which are left abandoned or seen as dangerous in most housing blocks.

These strategies reflect our ambition to prepare the ground for an ecological transition and make something out of this modern ground that meets the challenge of the present ecological and societal transition (Fig. 11.5). This approach is more metabolic than functionalist: it includes technical solutions and does not fear becoming transformative and experimental, but above all it aims to meet the requirements of human uses. This large public property represents an opportunity to create a 'Landscape of the Transition' that should not be missed. Nevertheless, the importance of a major figure in its maintenance, the gardener, should not be overlooked. Therefore, this transition must go hand in hand with residents' appropriation of and education on these new systems (ditches, late mowing) and public authorities' maintenance of these landscapes. This is essential to maintain the quality of a diversified landscape over time and slowly adapt practices and lifestyles.

11.7 Mixité: Emancipation and Initiatives

The ground of the city constitutes a common surface of interaction between the different projects or aspirations of the city's inhabitants. By means of an equitable distribution of programmes and accessibility to a plurality of social environments, urban space can encourage the possibilities for one person or more people to emancipate themselves through the development of individual or collective initiatives. However, the equity of these conditions is still not distributed to the same extent in the different parts of the city, and this common space contains asperities and omissions, so we must look for ways to avoid any further marginalization by questioning its future role.

The case of Peterbos, with nearly 3,500 residents, is no exception to the well-known imbalances found in large social housing estates. Urbanity is excluded there by isolation and spatial decay coveted by drug traffickers, gradually giving way to forms of territorial micro-tyranny and a gradual alteration of the neighbourhood's identity via the media. The late arrival of urban renewal policies in this type of neighbourhood and a certain denial of reality by public authorities has inevitably weakened social ties and the possibility of interaction with residents. They have in part become resigned and indifferent to any initiatives or participation developed by traditional planning policies, which often require a civic practice and a cultural codification difficult to apprehend (Pisecki 2007). Living in Peterbos literally means living in and with public space, and this is precisely what enables us to question our

Fig. 11.5 Setting the scene for the ecological transition

design practices and observe the limits to our ability to formulate a response beyond a mere reinterpretation of public space, namely interacting with specific practices and uses.

During the development of the master plan, we worked on the missing dimension of the 'intermediate urban grain' in Peterbos, which are spaces that are not part of the private sphere of housing and do not concern the large public space but can encourage more informal and small-scale practices, such as a repair workshop, an ephemeral coffee shop, and small-scale agricultural production and distribution. This idea is regularly considered in view of managerial, economic, or even regulatory difficulties, such as PRASS (*Plan régional d'affectation du sol*)-related restrictions on the possibility of using the ground floor of the buildings. Indeed, the latter is limited by the categorization of the different parts of Peterbos into predominantly residential areas and parks, as well as equipment and service areas for the collective or public interest, highly constrained by floor-surface limitations. This frustration is all the greater when you consider the number of jobseekers, the interest that emerged during discussions with residents in opening individual businesses, and the repercussions of inactivity for vulnerable people in Peterbos. It therefore seems fundamental to rethink on a regional scale the possibilities left to these places to reconceive this spatial intermediate grain and to undertake initiatives that meet specific needs and wishes. This missing intermediate grain is also applicable at a social level. Despite the large number of associations present in the neighbourhood, the social projects developed by public urban renewal policies, and the creation of a neighbourhood and participation branch in 2019, interaction between representatives and those represented is still very difficult. Moreover, one may question the means currently put in place by regional and municipal authorities to ensure the sustainability of these socio-economic projects—undertaken within the framework of its 'Sustainable Neighbourhood Contracts'—since they no longer ensure the follow-up of these projects after the construction operations.

Previously, life in these neighbourhoods revolved around the figure of the 'caretaker' concierge, a key intermediary and a great source of knowledge of the various situations. Although some of them remain in the western part of Peterbos, their roles and involvement in the life of the neighbourhood have been greatly reduced to that of a technical and sanitary manager of the buildings. In recent years, this need to 'take care' has been taken up by the *opbouwwerkers* of the association *Samenlevingsopbouw* or by the educators of the nonprofit organization D'Broej who, as its name suggests, try to 'deal with' the resources available (Fig. 11.6). These associations, along with SoHoLab, have been valuable sources for learning about issues and practices, particularly for the design of public space in the neighbourhood. However, it is obvious that these structures will not be able to resolve a situation on their own that arose out of the obsolescence of a territorial management policy for several years.

The Peterbos experience leads us to a better understanding of what is missing in the balanced implementation of programmatic distributions and social cohesion factors in marginalized parts of the city. This common plane of interaction is therefore the result of a long process of listening and monitoring initiatives that are often fragile at the start. Setting up a coffee shop, installing collective farming practices, providing

Coffee meeting in Peterfoot

Food distribution by the association Samenlevingsopbouw

Education of the youngest in beekeeping on the future agricultural plain

Conversation with young people about the use of sports areas in the neighbourhood

Fig. 11.6 Mixité: emancipation and individual initiative

mutual aid, and even organizing sports and cultural events are all aspirations that are difficult to initiate and perpetuate in Peterbos today. In addition to overcoming regulatory constraints and designing the intermediate grain of the space, it seems interesting to rethink the figure of the 'caretaker' (Fleury 2019) as an intermediary able to encourage and follow up on residents' projects. This figure could take the structure of a cooperative, facilitating the interaction and action of the actors involved in the longer term to support certain experimental initiatives, while ensuring a form of project management (Dimeglio and Zetlaoui-Léger 2007). The Peterbos experience reveals to each stakeholder, politician, owner, designer, academic, or association the need to find other configurations and construction processes for this social space, which cannot simply be the repetition of a 'daily problem-management' policy. The future of a *mixité* for these districts is therefore a project in which urban design participates as support and from which things are made possible, but whose results cannot be expected at a time 'T' but should rather be envisioned as a continuous process.

11.8 A Provisional Conclusion: The Stone Guest

During the 1960s, 1970s, and 1980s, urban debates in Brussels centred around the *luttes urbaines* (urban struggles). Under the term *bruxellization*, standing for market law and the real estate operation of 'modernizing' Brussels by intense housing demolition and expulsion, activists and residents denounced the lack of a public voice in the urban question and reclaimed a 'right to the city'. The end of the 1980s concluded with the creation of the Brussels Region, a regional institutional layer covering the nineteen communes. At the time, the causes and demands of the urban struggles culminated in the implementation of an official urban renewal policy, the main outcome of which is still the 'Sustainable Neighbourhood Contract' (Lenel 2013; Van Criekingen 2013). According to Van Criekingen (2013), the institutional birth of the Brussels Region in 1989 led to a project that favoured the return of the middle classes to the city, which went beyond the objective of social cohesion in working-class districts. Another element in the Brussels context of access to social housing relates to management by social housing companies, which moved from the municipal to the regional level in 1993. The change from an imperfect system, which left access to housing dependent on contacts and subjective assessment but implicitly favoured a form of village mix through the fluctuation of the selection process, to a more transparent system with the consequence of homogenization, concentration, and impoverishment of the inhabiting social strata. Furthermore, the departure of the middle classes who no longer fit into the redefined categories was experienced as class stigmatization by those who remained (Schaut 2017). Both of these elements suggest that disinvestment in social housing neighbourhoods dates back to the turn of the 1980s and 1990s and has worsened since, although many reinvestment programmes have been developed.

The radical transformation of Peterbos is now becoming a necessity and requires not only rethinking the social and physical context but also deconstructing the implicit socio-political project that stemmed from a series of likely good intentions, confronting fundamental questions on the place and role that we wish to grant these districts in our societies in the future. However, urban design and urban planning have major responsibilities in that they 'govern by design' (Aggregate 2012) and can worsen inequalities or revise them: 'any policy which seeks to eliminate or combat them must take the urban project as its starting point' (Secchi 2014). It is then important to note that the Peterbos Sustainable Neighbourhood Contract is the first to take over the entire perimeter of one of these neighbourhoods and is accompanied by actions that concern public space and housing quality as well as social interventions. The ground transformation we aim to initiate, with the multiple meanings developed here, will never bring welfare, social cohesion, inclusion, and jobs by itself. But with its insistence on space as capital, an agent, and a mediator, it tackles the stone guest: it clearly highlights the possible starting point for reform and the assumption of collective responsibility.

References

Aggregate (2012) Governing by design. University of Pittsburgh Press

Bourdieu P (1993) La misère du Monde. Seuil

Bruxelles Mobilité (Ed) (2020) Good Move. Plan régional de mobilité 2020–2030. Bruxelles Mobilité

Cahn L, Deligne C, Pons-Rotbardt N, Prignot N, Zimmer A, Zitouni B (2018) Terres des villes: Enquêtes potagères de Bruxelles aux premières saisons du 21è siècle. Éditions de l'éclat

Commune d'Anderlecht (1963) Anderlecht commune verte

Degros A, Decleene M (2014) Bruxelles, à la [re]conquête de ses espaces. L'espace public dans les contrats de quartiers durables, Service public régional de Bruxelles

Dimeglio P, Zetlaoui-Léger J (2007) Les rapports ambigus entre politiques et citoyens: Le cas du réaménagement du Quartier des Halles à Paris. Fr Polit Cult Soc 25(2):115–140

Fleury C (2019) Le soin est un humanisme (Vol 6). Gallimard

Leloutre G (2017) Le Park System d'Anderlecht. Construction d'un espace public pour la couronne verte bruxelloise. Bruxelles Patrimoines 22–23:114–129

Leloutre G (2020) La transformation moderne de Bruxelles. Processus d'agencement de l"espace urbain bruxellois entre 1949 et 1979 [Unpublished doctoral dissertation]. Université Libre de Bruxelles

Lenel E (2013) La mixité sociale dans l'action publique urbaine à Bruxelles: Projet ou langage politique? Brussels Studies 65:1129. https://doi.org/10.4000/brussels.1129

Pisecki GS (2007) La mixité urbaine à Bruxelles: Interprétation locale à travers le discours [Unpublished doctoral dissertation]. Université libre de Bruxelles

Schaut C (2017) Tout au Nord des tropiques: Une architecture moderne à Bruxelles. CLARA 4(1):203–218. https://doi.org/10.3917/clara.004.0203

Secchi B (1986) Progetto Di Suolo. Casabella 520(521):19–23

Secchi B (2014) La ville des riches et la ville des pauvres (M. Logoz, Trans.). MetisPresse

Smets M (1986) La Belgique ou la banlieue radieuse. In Fondation de l'architecture (Ed.), Paysages d'architectures (catalogue d'exposition). Bruxelles, pp 33–35

Sterken, S. (2013). Bruxelles, une capitale en mouvement? 50 ans d'architecture et d'urbanisme. In Bruxelles Patrimoines (Vol. Hors série, pp. 187–203). Bruxelles: Région Bruxelles Capitale - Direction des Monuments et Sites.

Strauven, F. (1983). Renaat Braem, architect. Bruxelles: Archief voor Moderne Architectuur (AMA).

Studio Paola Viganò, & vvv architecture urbanisme (2020) Masterplan Peterbos. Bruxelles.

Van Criekingen M (2013) La gentrification mise en politiques. Métropoles 13:4753. https://doi.org/10.4000/metropoles.4753

Paola Viganò architect and urban designer, is Professor of Urban Theory and Urban Design at EPFL (Lausanne) and IUAV Venice. She was *Doctor Honoris Causa* at UCL in 2016, *Grand Prix de l'Urbanisme* in 2013, and received the Flemish Culture Award for Architecture in 2017 and the Golden Medal for her career at the Milan Triennale in 2018. In 1990, she founded *Studio* with Bernardo Secchi, working on projects and visions in Europe. Since 2015, *Studio Paola Viganò* has won several international competitions and works mainly on public spaces and urban and landscape projects.

Accomplished projects by *Studio* include the Theatre Square and Spoornoord Park in Antwerp, both awarded prizes for the best public spaces in Flanders, and the City Structural Plan; a system of public spaces in Mechelen, finalist of the European Prize of public spaces; the Courrouze district in Rennes, which won the Prix 'Nature en ville' of the Palmares Ecoquartier 2011 awarded by the Ministry of Ecology in France. Today, *Studio* is working on the projects Brest metropolitan Heart, Peterbos public spaces in Brussels, the Flaminio neighbourhood in Rome, and Lecco waterfront, among others.

Bertrand Plewinski graduated in architecture in 2014 and obtained an advanced master's degree in urban and regional planning in 2017, both at the Faculty of Architecture, and Architectural Engineering and Urbanism of the UCLouvain. He acquired professional experience at *Studio Paola Viganò*, where he has been practicing since 2016. He has worked on several territorial studies and master plans in Belgium and France (Brest Metropolis, the guide plan of Val-de-Sambre, and the Urban Renewal Contract for the Citroën-Bassin Vergote-Parc Maximilien). Currently, he is coordinating public space projects in Brussels (Peterbos Park, Marie Janson Square) and the follow-up to several competitions in Belgium.

Guillaume Vanneste graduated as an Engineer Architect from EPL UCLouvain in 2009 and is a researcher and a visiting lecturer at the Faculty of Architecture, Architectural Engineering and Urban Planning of the Université Catholique de Louvain (LOCI-UCLouvain). He recently defended a PhD thesis which explores the production of the contemporary city-territory landscape and land structure in Belgian dispersed urbanization. Exploring the case of Brabant from the perspective of land ownership and land structure has deepened the understanding of contemporary urbanism in Belgium. He acquired professional experience at Studio Secchi Viganò and is a founding partner of vvv architecture urbanisme in Brussels.

Nicolas Willemet is an architectural engineer and took part in teaching activities and academic research projects on public space in the metropolitan region of Charleroi at the Faculty of Architecture, Architectural Engineering and Urbanism of the Université Catholique de Louvain (LOCI-UCLouvain). He is a founding partner of vvv architecture urbanisme in Brussels and teacher at ULiege. Over the course of his career, he has gained experience in Switzerland and Belgium in renowned architecture offices such as Localarchitecture and L'Escaut.

Within vvv, Guillaume Vanneste and Nicolas Willemet develop architectural projects and urban design in Brussels and Belgium. In particular, vvv has won a number of public projects, such as the Pocket parks and the C28 promenade on a former railway, the renovation of Place Marie Janson in Saint-Gilles, and the master plan for the Peterbos Housing Park in Anderlecht. The office's approach is to conceive spatial structures with care and interest in social and material terms while always maintaining a critical research perspective.

Chapter 12
Participation and the Architect: Creative Partnership or Communication Breakdown?

Dominique Lefrançois

Abstract Studies on participation tend to focus on describing the process. The exact form of their involvement, along with its effects on architecture and the lived environment, is rarely discussed. The aim of this article is to better understand the effects of participation in the long run after the initial involvement. This will be done by studying three rehabilitation projects in the Paris region. Our aim was to gather the opinions of those who had not made their voices heard through the official channels of consultation. Above all, it was to give visibility to, and understand the meanings of, the appropriation of spaces which have become invisible simply because they are used in ways which seem to go against the norm.

Keywords Ethnography · Public space · Participation · Urban ecology

12.1 Introduction

Since the very beginning of urban regeneration policies in France in the 1980s, a recurrent criticism levelled against the forms of town planning and architecture inherited from the '*trente glorieuses*'[1] has been that residents find it difficult to take ownership of these spaces. The refurbishment of large social housing estates is often based on the notion of bringing them into line with the norms of what a 'city' is considered to be, namely a compact space following the model of the historic city centre, rebuilt successively within the same footprint (Vayssière 1988). Yet, is this not a rather narrow vision of urbanity? Many districts once situated on the periphery have been so completely swallowed up by expanding cities that are now an integral part of new, enlarged urban centres.

[1] The three decades of postwar economic growth.

D. Lefrançois (✉)
Unité de Recherche AUSser, Ecole Nationale Supérieure d'architecture de Normandie, Rouen, France
e-mail: dominiquelefrancois4@gmail.com; dominique.lefrancois@rouen.archi.fr

N. Aernouts et al. (eds.), *Urban Living Lab for Local Regeneration*,
The Urban Book Series, https://doi.org/10.1007/978-3-031-19748-2_12

Here, neighbourhoods are still regarded as enclaves, their inhabitants discussed solely in terms of crime and welfare dependency, as if they were untouched by developments in the way people live, and in particular the way in which the delineation of urban and residential districts alike can move, or *be made to* move (Estebe 2015; Lepoutre 1997). Might observing the social practices of these same residents and their involvement in architectural and urban development projects not help enrich the way in which we regard the large out-of-town estates of the French *banlieues* and move away from the stigmatizing—and highly distorted—image traditionally applied to them?[2]

This, then, is the central question of the ethnographic study presented in the outline here. By revisiting three refurbishment projects carried out in social housing estates around Paris, all based on the principle of resident participation, it also evaluates the usefulness of participation, an integral part of French urban regeneration policy since the 1980s (Bacqué and Sintomer 2011). Much has been said and written about participation in France: desired and reviled in equal measure by public actors, architects, social housing providers, commissioning clients, and local officials, it was paid considerable attention by French researchers in the 1970s and again in the 1990s and, to a greater extent, the 2000s (Blondiaux and Fourniau 2011). The researchers in question were generally at pains to show how these supposedly participatory projects were, in practice, not very participatory at all. It is true that these projects[3] involved only a small number of people, and those who did express their opinions—which tended to reflect the preoccupations of the middle-class and/or retired population—often did so at the expense of other groups (young people, immigrant populations, etc.) who were largely absent from consultation sessions. The aim of opening up deprived areas to new populations, and the French government's policy of social diversity—which seeks to attract the middle classes to these districts—also have their part to play, in this diagnosis.

Above all, as sociologist Rosenvallon (2020) passionately argues, there is a need to find ways of consolidating what is a rather 'elastic' concept of participation and reconfiguring it to respond to the practical realities of a plural, post-electoral, post-majority society; to find ways, more generally, of representing the governed as a whole, and not just those supposed to represent a majority of people—a majority which no longer sees itself as being reflected through these intermediaries.

The crisis in our democracies stems from the fact that society no longer feels represented in its true diversity (Rosenvallon 2020). Local officials, commissioning clients, and social and urban actors still tend to see society in terms of large blocs or categories—professions, trades, and social classes—even when the individuals in question no longer feel that they belong to a group and do not define themselves in terms of stable social practices but rather in terms of trajectories, of constant disruption, of communities of experience, rather than of belonging, all of which

[2] The term *banlieues*, or suburbs, designates the areas dominated by public/social housing found on the outskirts of France's major cities and has become a byword for social deprivation and exclusion.

[3] The same criticisms apply to the various forms of participation which could have involved residents earlier in the planning phase, in particular regarding the design of furniture and other equipment.

are in a state of flux. Rosenvallon thus calls for the development of a form of non-electoral representation. For, in his view, to build a society one must first build an understanding of how people live, and in so doing put an end to the unawareness that we have of one another. From this comes the need to construct a new narrative, to give expression and public recognition to the opinions and the lives of all. The present article, which deliberately mixes descriptions of social practices and statements from residents with scholarly analysis, thus stems from this desire to provide an account of the life of places and the people who use them.

The question here is a political one. One has to want to listen to people and know how to do so, taking account of the diversity of what they say, what they are reluctant to say, and what they desire. The architect Andrei Feraru, who was in charge of the three post-refurbishment projects examined here, certainly sought to do this. Following his sudden death during our research, we pay tribute to him here, as not only an architect but also as a man of the theatre who in his spare time explored various ways of letting people speak, through the narration of experiences and stories alike. Might artists and actors be able to function as intermediaries to bring out the desires, needs, and practices that not everyone is able to express, and to make these voices heard?[4]

In 2006, the approach Feraru chose for this refurbishment was firmly participatory, and we will therefore examine his method here by returning to the three social housing estates involved: the Cité de la Fosse aux Loups in Fresnes, the Villa Saint Martin in Longjumeau, and the Jean Bouin residential complex in Taverny, all property of social housing provider (hereafter SHP) 3F, which employed him along with a landscape designer, Armelle Claude, and sociologist Véronique Blin from the sociological consultancy company GERS. Residents were invited to take part in participatory workshops led by the sociologist, who was also tasked with conducting a study into residents' social practices. The issue of sustainable development was also raised as a particular priority, given that the mayor of Longjumeau, Nathalie Kosciusko-Morizet, had just been appointed to the government as Minister of Ecology.

[4] Working with his wife, Patricia, Andrei Feraru cowrote plays, including *Making Off*, performed at the Théâtre de l'Opprimé in Paris, a work based on the 'forum theatre' model, which seeks to make social and political conflicts visible by giving voice to marginalized groups. This work was to have been performed at the Maison de la Culture in Genevilliers, the town in the Paris *banlieue* where Andrei Feraru was overseeing the refurbishment of a large estate. Taking the form of a large debate, it presented the creative process involved in his forthcoming play *12 jeunes en colère* (*12 Angry Youths*) based on Sidney Lumet's film. Written collectively by a group of authors and performed in several languages (French, English, Italian, and Romanian), the play shows how immigrants live in one language while dreaming or working in another and features characters from different backgrounds: Chinese, Italian, and Algerian. More generally, it examines how a society regards its young people, who are seen as listless and insecure.

12.2 A Resolutely Plural Approach to Collecting Opinions: An Ethnographic Method

The present study presents the results of an investigation carried out on the ground with Mina Saidi, an architect and ethnologist involved in ongoing research at the School of Architecture at Paris La Villette and the Normandy School of Architecture. It also builds on existing studies of Living Labs conducted with Nadya Labyed (Sevran Cycloterre project, Caen-La Mer). We have sought to assess how the refurbishment schemes were received by residents—whose reactions could range from appreciation and appropriation to criticism and outright misuse—and the extent to which these corresponded (or not) to residents' needs and varied practices.

Our aim in returning to these three sites was to gather the opinions of those who had not made their voices heard through the official channels of consultation. A neighbourhood is defined not only by its inhabitants but also by their movements—those who live in it, work in it, or simply pass through it. Consultation is often envisaged in geographically limited terms or focused on people thought to represent a neighbourhood. However, today's neighbourhoods stretch beyond their strict geographical boundaries.

The acknowledgement of disparities in people's ability to express their views in public (Bacqué 2015; Bacqué and Mechmache 2013; Blondiaux and Fourniau 2011; Zetlaoui 2013) underlines the need to engage with a wider variety of populations and also adopt other modes of data gathering—in addition to straightforward listening, for instance, the observation of places and gestures. The ethnographic method that we employed involves intensive data gathering on the ground, observing these areas at different times of the day and also over time, for a period of six months. Our aim was above all to give visibility to, and understand the meanings and modes of, the appropriation of spaces which have become invisible simply because they are used in ways which seem to go against the norm, in spite of their contribution to the upkeep of the neighbourhood. Studies on participation tend to focus on describing the process, the degree to which people become involved—or not. The exact form of their involvement, along with its effects on architecture and the lived environment, is rarely discussed. Just as 'guerrilla' acts and practices are ignored, in many cases the voices of residents are not heard, as they seem to be contrary to a fixed idea of what cities are, even if in reality these are constantly evolving. Far from being a constraint, as one often hears in this profession, might the involvement of residents not actually help open our eyes? And fuel—rather than, as some complain, stifle—the imaginations of architects and their commissioning clients?

It is, however, important to note that participation by residents tends to seek consensus, and as such is not always representative of the diversity of individual opinions. Consequently, it generally leads to highly normative responses; in other words, the exact opposite of what the Living Lab method analysed in this book—which suggests the idea of experimentation, the search for new ideas and ways of doing things—is expected to produce.

12.2.1 *Plan*

First, we examine the question of public space. In this respect, the refurbishment can be considered a success insofar as it has enabled the appropriation of this space by residents. However, the way this has occurred has worried building managers and caretakers, in spite of the fact that they had themselves asked for these spaces to be used. A nice smooth, clean, empty floor may appear easier to maintain. Yet what could be more participatory than seeing people engaged in activities in public spaces? Is it not the case that public space is the site of arguments, of debates, and of meetings more generally? This is how we have experienced it at any rate. Our status as outsiders in the neighbourhoods under study, along with our willingness to bring together people of different ages and belonging to different categories who rarely feel able to speak to each other or are even used to treating each other with outright hostility—youngsters and older people, social housing tenants, and neighbouring homeowners—has allowed people not only to express their views but also to enter into dialogue with each other.

Second, we examine the way in which ecological imperatives have guided refurbishment works. Here, the verdict of residents has been less favourable. One might even ask whether in this case participation brought about an effect opposite to that hoped for: at the end of the process, residents felt that their requests were not listened to, even though their opinion was sought, leading them to disengage from these participatory mechanisms. This is in spite of the fact that they may have clear ideas on this subject, which come from their experience of a lived urban space different from that of the city-centre model, constantly presented to them as an ideal. Shifts in lifestyle and the environmental question are shaking up many boundaries, be they territorial or disciplinary, including those of this project. With participation still very much on the agenda, then, it is important to turn our attention towards the places in which it takes form.

12.3 Spaces Open to All, and no Longer Dominated by Men

What were these planned improvements, then, and how have they been received? It was the intention of the designers and architects, let us not forget, to lure people into outdoor spaces by employing architectural legerdemain. At Longjumeau, this involved creating a central park where there had been a patch of scrubby grass, along with a shared biodiversity garden. On all three estates, children's playgrounds were positioned not far from the main gates, and the rather shabby lawns were spruced up. A frequent criticism of estates like these is that the public spaces are dominated by a certain age group, and more generally by men. The media tends to focus on young people left to 'fend for themselves' who, the argument goes, have completely taken over public spaces. As a result, some sociologists conclude that these spaces are avoided by other people (Donzelot 2012). Another worry, expressed by the mayor of Fresnes, relates to the tendency towards isolation and increased loneliness among residents. The architect, who was concerned about the small size of the apartments, saw the focus on improving outdoor areas as a chance to provide residents with added living space.

Today, the improvements seem to have largely fulfilled these objectives. At Longjumeau, the park and playgrounds have proved popular, and are used by large numbers of people. The fact that parents are happy to let their children play outside is seen by 3F as proof that the new improvements have brought a new sense of security. Indeed, children stay outside until late at night, much to the annoyance of the caretaker, whose job it is to keep these public areas clean and orderly. This practice does go against convention, which holds that children should be in bed by a certain time. The caretaker sees the presence of children at what is judged an unreasonable hour as proof of the well-worn accusations levelled at supposedly negligent parents unwilling to look after their offspring.

Overall, then, the use made of these refurbished spaces corresponds to what the architect wanted and expected. Public space, in the form of playgrounds and a park, constitutes an extension of the home by virtue of being regularly used. These have become predominantly female spaces, serving as living rooms for women who, rather than inviting them indoors, meet other women in these extensions to their apartments. For example, a daughter who has moved to another estate where she does not feel a close connection with her neighbours knows that here she can find her mother and her old friends from her apartment complex.

Given the large number of women working as childminders on these estates, public squares and parks also appear to fulfil the role of workspaces. This type of work is accessible to women without qualifications living on the estates and can be done at home. Home, however, is a small place. Thus, in order not to bother their own families and to let the children they look after run around, many women exploit these outdoor spaces.

Might it also be the case that a lack of high-quality parks elsewhere brings in people from surrounding areas? Among the people from other neighbourhoods who frequent the park on the Saint Martin estate in Longjumeau, there are not only childminders but also grandmothers looking after their grandchildren. This was the case for one woman living in the neighbouring condominium. Her own apartment complex, now occupied by an ageing population, has removed its playground which, for lack of children, is now just an ornamental area that no one is allowed to walk on. This property owner is therefore in the habit of bringing her own grandchildren to play in the playground and park of the estate next door. Indeed, one young man we met a short distance away joined our conversation, telling us that ‘the great thing about the estates is that there are all these young people and children (there are lots of families, both French and foreign!)’. The lively atmosphere found here benefits the neighbouring complexes, home to a wealthier but older population whose grandchildren have school friends who are allowed to play outside the apartment buildings, as well as college students, who sit on the benches at lunchtime while the college gates are locked, preventing them from gathering outside.

12.3.1 Social Diversity: An Objective Achieved More Through Public Spaces Than Housing

Are the neighbourhoods which today sit protected behind their gates thus profiting from the existence of more open social housing estates? SHP 3F wanted to move away from the model of 'residentialization', a form of refurbishment which in the 1990s–2000s saw estates being split into a multitude of residential units fenced off from their surroundings in order to give the impression of middle-class apartment complexes. Forty years of French urban regeneration policy appear in any case to have borne fruit: the money invested in these estates has by no means been wasted. On the contrary, one reason why a certain number of residents do not wish to leave these estates is that they now boast facilities which are not found elsewhere. As far as the caretaker is concerned, however, the appropriation of these spaces is a failure. He explains that he has the job of dealing with the extra rubbish generated by these crowds of people. The bins are overflowing because they were designed to be used by residents only. Since they are being asked to take on an ever-growing list of administrative duties, caretakers are unwilling to deal with the consequences of the refurbishment programme. As one of the caretakers states: 'The estate is open, it serves the needs of all citizens, the problem is that its upkeep is down to us in our apartment complex.' Indeed, should the local council not assist in managing and financing a public space which is situated within a private apartment complex owned by 3F, but which has been transformed through its use by the public into an urban park?

For this caretaker, the park is an eminently public area, given the diversity of people encountered there. In contrast to Donzelot's concerns about the lack of public space in such neighbourhoods, another sociologist, Remy (2015), draws a distinction between residential public space and public space in city centres: the latter is more anonymous, open, and passed through by all and therefore resists occupation by any one individual or group rather than another (Habermas 1988; Joseph 1992). In a residential neighbourhood, by contrast, public space can be occupied for long periods without it necessarily bothering others, for the simple reason that rules of good neighbourliness are applied here. Women, young people, and children, for instance, who are all attracted to the same space, take turns occupying it at different times.

However, the park at Longjumeau is large enough to allow for the copresence of individuals who do not wish to speak to each other. It fulfils the role of a central public space, being frequented by women from neighbouring estates (Longjumeau is surrounded by several other social housing estates and private complexes) even though relations between these estates are tense, if not openly hostile. It is also used by women and men looking for a bit of solitude. Its size allows people to do something that is often described as a problem on these estates, namely cut themselves off. Madame G., for instance, comes down when she wants to read or get some peace and quiet. A mother of seven children is able to find a degree of privacy here that she cannot always get at home.

12.4 Ornamental Principles Out of Tune with the Lived Environment They Are Intended to Enhance

There are criticisms, but these mainly relate to the designers' overly normative conception of the facilities. 'Are these architects dwarves?' exclaims Madame C. on La Fosse aux Loups estate. This mother considers the playground equipment to be ridiculously small, as she sometimes needs to help her small child get into the Wendy house and would like to be able to join the game herself. Developers consider playgrounds to serve only one purpose—play—and only one age group: very small children. The playground is also designed for children of a specific age. Yet playgrounds are like clothes: children are constantly growing out of them. The equipment here is unsuitable for children who might like to play there when they are older.

Young children are better catered for than teenagers, who are very critical of the refurbishment. The spaces provided for the latter are poorly designed (such as a football pitch placed on a slope). Situated on the outskirts of the estate, hidden away, they are not used much. Indeed, young people are seen less as a specific age group with their own desires and problems and more as troublemakers (Mauger 2019). The improvement works have sought above all to reduce their presence. In Fresnes, the central space now occupied by the children's playground has pushed them to the outskirts of the estate in order to increase a sense of residential privacy. As a result, footballs have constantly been landing in the garden of a private house adjacent to the estate, giving rise to real conflict, first between the young people and the owners of the house and then between the latter and the children's parents. This dispute resulted in the building of a wall separating the house from the estate; this was paid for by the couple who own the property, as 3F did not respond to any of their complaints.

Does participation mean that adolescents lose out, thus exacerbating intergenerational conflict? Is this a consequence of an increased feeling of insecurity?

While young people may have contradictory impulses, making it difficult to keep them happy, so too do their elders. The latter may feel worried at seeing them continually hanging around in front of the buildings, but at the same time reassured to find them there when they come home late at night or laden with shopping, knowing that there will always be someone to help them if need be. This mix of ages enables a sort of informal building supervision to be maintained. The variety of outdoor furniture also helps with this, directly contradicting the theory of situational prevention, which states that a clear layout of facilities focused on a single function is needed to prevent conflict. Thus, while children are left outside alone, this is only because adults can rely on the informal supervision of their children by other residents, who are keeping an eye on their own children from their window or at the playground. Forms of sociability associated with inward-looking communities are thus not incompatible with more informal types of solidarity.

When it comes to the subject of sustainable development, more strident criticisms are voiced. Residents are unwilling to adhere to ecological measures and, fifteen years on, the shared garden at Longjumeau is abandoned and in a sorry state. People abhors a vacuum, as preventive architecture tells us, and the marsh areas filled with rubbish confirm this. As one woman explains, 'we haven't necessarily got green fingers, so we don't care about the gardens'.

Residents have more pressing concerns: instead of marsh areas, they would rather see parking spaces which they are continually asking for, or areas for barbecues. Indeed, they are dubious about the usefulness of these environmental measures, which they see as pointless. All the more so, because they have been left disappointed by broken promises: the adoption of eco-measures has not led to financial savings for households (Ortar and Subrimon 2018). Rising energy prices and the individualization of energy bills in particular have contributed to weakening the position of households, even though they have been trying to comply with this discourse. Cars, meanwhile, which the refurbishment scheme has sought to exclude as much as possible, are an absolute necessity for work in the suburbs and also make things easier for families, when grown-up children end up having to live with their parents for long periods.

These ecological measures are driven by educational and moral, even paternalistic objectives, which tends to put people off them. The recycling points and shared gardens that were supposed to act as showpieces have turned out to be anything but. Promoted today as a way of introducing biodiversity to urban areas, in the 1990s gardens were touted as a method of creating social interaction in deprived areas. Then, when concerns about crime came to the fore, it was hoped that they would help reduce feelings of insecurity by encouraging residents to take ownership of their space: it was imagined that a shared attraction to gardening would overcome neighbourhood conflicts. The department in charge of participation at 3F sees gardening as a socialization tool, bringing people together, fostering greater cooperation in local affairs, and encouraging tidiness. For us, this is a questionable approach. These measures do not take into account the practices of residents, who are largely unconcerned with ecological imperatives. They also fail to acknowledge differences in taste with regard to the built environment, which have already been analysed by ethnologists (Depaule 2004; Dubost 1999). When lawns are replaced by rampant weeds, this rather detracts from the desired image of a well-ordered estate.

The newly installed bins are technological marvels. Yet, by ignoring social practices, they risk being misused, misunderstood, and ineffective, overflowing as they are with litter and surrounded by bulky objects piled up beside them. Their unsightly state is blamed on people's incivility, yet for us it provides evidence to the contrary. For, if the bins are overflowing, this is down to various reasons. One is brought to our attention by the caretakers, who are angry with parents who let their children take out their rubbish bags: 'The new bins are too high or too tightly closed and therefore inaccessible to children'. They throw the bag in, without being able to close the lid of the bin, and consequently the rubbish goes all over the place. But what is seen here as irresponsibility can also be read as a sign of civic responsibility and an equitable division of household tasks, whereby children run small errands (buying bread, taking out the rubbish). It is evidence of how children are being taught to join in shared efforts and help with daily tasks.

The bins are also overflowing because no thought has been given to an urban practice which is widespread in our cities, but particularly well developed in estates like these, and akin to a form of recycling. Many people rummage through the bins, some looking for food, others for clothes or furniture. Some trade the items they find on the Internet or markets and get their stock from the bins, where one can find bags, old shelves, shoes, or appliances in working order. These overflowing bins are also a sign that people have to move frequently. But these items are thrown out on the understanding that they could be used by someone else. Mattresses and bags of usable clothing will therefore not be placed in the bin, but next to it.

12.4.1 Has the Original Layout of Major Estates Lasted Better Than today's Fragmented Refurbishment Projects?

One criticism of the outdoor spaces on estates built in the 1950s–1970s is that they neglect the spaces next to buildings and pedestrian use more generally, concentrating instead on the view that residents have from their windows (Sennett 1992). Yet, is this not one of the better points of these estates, with their tower blocks and massive high-rise units? Designed on a scale to fit into the 'grand landscape', these offer magnificent outside views. The remarkable sites chosen for many estates—on hillsides or at the edge of forests, as seen in Marseille, Paris, Rouen, and Nancy—are another element in their favour. In this sense, have these colossal *grands ensembles* estates perhaps lasted better due to the consideration given to the wider landscape in this period? Today, by contrast, through the residentialization policy and the diversity of planning instruments, design tasks are compartmentalized and sites are split into plots, forcing landscape architects to work on a much smaller scale.

Critiques of modernity have also led designers to the model of separate plots, reimposing it on a public space originally envisioned as being free from the limits of private property, stretching out into wide horizons. However, is it not time now to rethink the boundary between the built environment and public space, but also between humans and the natural world, between housing and the landscape? If refurbishment schemes seek to foster better relations between people, could they not also find better ways for humans to coexist with animals? One of the reasons why residents are unhappy with their biodiversity gardens is that these increase the number of bees and other insects that they are less willing to accept, while the SHP is unable to eradicate other pests—rats, bedbugs, and cockroaches—that plague the estates. Dogs are another source of conflict between neighbours. The association which looks after the shared garden, for instance, complains that dogs are always digging up their plants.

Could we get some ideas from residents who have, in fact, had one foot in the countryside for a long time? The wild boar roaming the forest edge in the Hauts Rouen neighbourhood, the childhoods spent birdwatching in Peterbos in Belgium, the housing blocks surrounded by cows at Sarcelles: these are all symbols of an urban environment situated on the edge of cities, which illustrate, as philosophers and ethnologists remind us, that the boundary separating the city from nature is a tenuous one (Descola and Ingold 2014; Desprez 2019; Latour 2015).

The names of these estates are also evocative of their surroundings. The *Cité du Renard* (the Fox Estate) in Fresnes leaves one female resident wondering whether, since the spread of ticks is linked to the disappearance of foxes, the latter could not be re-introduced into cities to emulate the situation in London, where they coexist alongside humans. As part of a workshop bringing together sociology and fine art at the Normandy School of Architecture, students curated a list of old recipes proving that, in the past, certain forms of coexistence between humans and other creatures were possible; and why not imagine animals living in 'mini-apartment blocks', which are already being sold for domesticated bees and other small animals?

12.4.2 Grey Areas: Spaces Which Are Neither Public nor Private, Promoting Insular Practices

As for the boundaries between public and private spheres, these are not very clear, as they vary according to both people's desire for individuation and the degree of intrusion of work into home life. This implies that living space, seen in terms which go beyond the physical boundaries of its walls, can include spaces of 'secondarity'—to use the term so dear to the sociologist Remy (2015): spaces where the individual can give free rein to their need for privacy and express their own identity. These would include the playground or the car park, the former functioning as a living room or meeting place for women, the latter being a more masculine space, serving as a workshop for DIY or mechanical repairs. They mark a renewed differentiation of the occupation of space along gendered lines, and at the same time signal the return of work into the private sphere of the home. Herein also lies the key to improving social relations in these neighbourhoods: the conflicts in question are as likely to arise inside the apartments as they are outside. Yet if the boundaries between spaces are blurred, so too are those between activities. Is doing DIY or mechanical repairs work or is it a leisure activity? Working as a childminder has much in common with the domestic sphere. DIY allows one to turn a space built or owned by someone else into one's own. It corresponds to Arendt's (1961) definition of 'work' (as opposed to biologically essential 'labour'): an activity which befits person and restores an identity which may have been left battered by unemployment or a low-grade job.

Public space, then, accommodates different degrees of publicness. It is therefore up to architects and those commissioning them to give form to these: legal specialists could also become involved, given that what is now looked for is a third way between ownership and renting, between public and private. The notion of third places or commons is back on the agenda (Garnier and Zimmermann 2018; Ost 2012). It refers to the shared use of 'common' spaces, an idea often dismissed as unrealistic in the context of deprived neighbourhoods owing to the enduring misconception that people from low-income backgrounds are unable to manage their own affairs. Yet, does the car park/workshop/maintenance area not already possess some characteristics of a third place or commons? This is certainly the case as far as its form and use are concerned (Lefrançois 2014).

12.4.3 Other Forms of Interaction Do Occur, but in Marginal, Liminal, or In-Between Spaces

For the moment, the community gardens are the only shared spaces to have been accepted. However, they have not proved popular in these neighbourhoods, as they seem to correspond to the desires of better-off populations, middle-class architects, and housing managers, whose jobs—particularly in the Internet age—are increasingly dematerialized, leading them to seek a return to the soil and manual tasks.

Instead of the participative schemes and Living Labs promoted on these estates, might these gardens go some way towards bringing about the social diversity that has been sought, unsuccessfully, through the construction of small apartments to attract middle-class residents? Diversity, in our view, has been achieved more effectively through the facilities offered than through housing. Having people from very different backgrounds living in close proximity in the same apartment complex generates mistrust and even conflict instead of the emulation and mutual understanding that was expected (Chamboredon and Lemaire 1970; Donzelot 2012). In this sense, the community gardens, like public parks, fulfil the role of public spaces in the sense that Habermas gives to this term: they allow a variety of people to meet, namely residents of the estates and populations from other neighbourhoods and social backgrounds. In residential areas, however, such contact seems possible on the margins only, in liminal or in-between spaces, in a slow and gradual manner.

The clearest example of such liminal, threshold spaces are the local shops. These benefit from their peripheral location on the ground floor of the apartment complex, facing onto the road which forms the boundary with other neighbourhoods. Located on this threshold, these shops have become true spaces of public expression and even of political discourse, to judge from the conversations overheard there, in which varied points of view, often defying stereotypes, are expressed by all on subjects such as crime, young people… The shops on the estate—a bakery with a terrace, a tobacconist and newsagent, and a pharmacy—thus function as real public spaces. The shopkeepers are simultaneously outsiders and intimately familiar with their surroundings, and as such well suited to the role of mediators (Tarrius 2002). Here, exchanges can occur between residents who are on bad terms or who normally find it hard to talk: between new and former residents, between homeowners from neighbouring areas and young people from the estate. It was in the tobacconist's shop that the 'mini-war' between the young people from the Fosse aux Loups estate and the couple who own the house next door was resolved. It was also the place where the residents of the new development and those living on the estate were able to stop eyeing each other suspiciously.

Is this the price at which the ecological imperative comes? That of acknowledging spaces of discussion—such as local shops—which let new and existing populations meet and allow urban density to gain acceptance? They are surely more effective in this respect than consultation committees, which do not attract many participants and even contribute to reinforcing conflicts—not only between young and not-so-young people but also between the working-class population and the middle classes, local officials, architects, and SHPs.

12.5 Conclusion: Taking a Bolder Approach to Interdisciplinarity and Breaking Down Barriers Between Professions

The problem is, in fact, that residents are constantly expressing their views—sending letters, signing petitions, writing to the press—yet are scared or simply unwilling to attend the consultation instances where a pretence is made of listening to them. Instead of the work on building interiors that they ask for, residents get a refurbishment of outdoor spaces. And similarly, while their involvement is sought in designing a bench through Living Labs and other workshops, they are forbidden from doing DIY in their apartments or in the car park. All the while, residents complain incessantly about electrical, plumbing, and heating problems, and about the walls painted on the cheap, with old materials, that are already looking shabby.

Might it not therefore be useful to prioritize dialogue with the actual actors involved in these projects in order to improve the quality of the services delivered? Social housing is a public service with a limited budget. Even so, is the problem here not more one of a lack of coordination between the various professionals involved in this work and the low quality of the services provided by these companies, which all belong to huge conglomerates? Driven by the sole objective of profitability, these firms use a precarious workforce made up of poorly paid casual staff (Jounin 2009) employed at day rates.

Might it not be productive to open up these discussions to a wider range of actors, including these companies, as the Caen la Mer Living Lab[5] plans to do? And why not include these undervalued workers themselves, re-imagining maintenance works as a space for exchange between residents and workers, as the architect Patrick Bouchain (2006) proposes? This would give new meaning to construction jobs that have become devalued because they have been dehumanized and increase the involvement of residents in decisions affecting their own homes. It would also help resolve another conflict, that between SHPs and their tenants, the former accusing the latter of being

[5] According to a project manager from the Caen la Mer social housing office, in the city of Caen whom we interviewed.

welfare-dependent whiners while residents accuse their landlords of being concerned merely with appearances and not with making material improvements.

Is it not right to involve residents in decisions on issues which concern them directly, such as whether cars are really as contrary to ecological principles as is supposed? Indeed, might residents be a source of innovation? For instance, the car park, currently seen purely in quantitative terms, could be seen in a more qualitative way as a space in its own right. Living Labs themselves—seen as structured spaces for meetings between different populations—could put themselves forward as spaces for ecological experimentation, challenging accepted ideas and offering a vision of these estates which goes beyond the city/nature opposition, as their residents, living as they do on the very edge of the city, have already understood. They could also find ways to reconcile the discrepancy between long-term accommodation and short-term social housing—meant to be a springboard for families who are not supposed to stay there for long but end up being unable to move into any other form of housing—through an acceptance of the idea that tenants should be able to adapt their accommodation a little more to their lifestyle, given that they will be staying for some time.

Ultimately, there is a need to reconsider existing norms, as the ecological imperative necessitates a different way of seeing the world. Yet participatory schemes and Living Lab projects are often too normative. Driven by operational objectives, they lose sight of the fact that, before one acts, one can question everything, and above all, that one should learn to look at things carefully. This can be achieved by teaching the rudiments of ethnology in architecture schools, not with a view to promoting the discipline, but in order to teach students how to listen, which is to say, not to confuse one's own desires with those of others. Existing spaces of exchange could also be taken into account, while the composition of teams working on these projects could reflect a bolder approach to interdisciplinarity. In addition to architects and landscapers, why not also bring biologists and legal experts on board, to add some fresh perspectives?

References

Arendt H (1961) La condition de l'homme moderne. Calmann-Lévy

Bacqué MH, Mechmache M (2013) Pour une réforme radicale de la politique de la ville. Citoyenneté et pouvoir d'agir dans les quartiers populaires; Rapport au Ministre délégué chargé de la ville

Bacqué MH, Sintomer Y (2011) La démocratie participative, histoire et généalogie, Editions La Découverte

Blondiaux L, Fourniau JM (2011) Un bilan des recherches sur la participation du public en démocratie: Beaucoup de bruit pour rien? Participations 1(1):8–35

Bouchain P (2006) Construire autrement. Actes Sud

Chamboredon JC, Lemaire M (1970) Proximité sociale et distance sociale. Les grands ensembles et leur peuplement. Revue française de sociologie, XI, 3–33

Depaule JC (2004) Savoirs et manières de faire architecturaux : populaires versus savants. Savant/populaire, Les cahiers de la recherche architecturale et urbaine, 15/16

Descola P, Ingold T (2014) Être au monde. Quelle expérience commune? Presses Universitaires de Lyon

Desprez V (2019) Habiter en oiseau, Actes Sud
Donzelot J (2012) A quoi sert la rénovation urbaine? PUF
Dubost F(1999) La maison, le beau et la mode. Terrain n°32, pp55–66
Estebe P (2015) La politique de la ville à la bonne échelle? L'introuvable solidarité d'agglomération. In: Kirszbaum T (ed) En finir avec les banlieues? Le désenchantement de la politique de la ville. Editions de l'Aube, pp 117–129
Garnier J, Zimmermann JB (2018) Solidarité sociale et proximités. De l'Etat providence aux communs sociaux. Espaces et sociétés 4(275):19–33
Habermas J (1988) L'espace public. Archéologie de la publicité comme dimension constitutive de la société bourgeoise, Payot
Joseph I (1992) L'espace public comme lieu de l'action. Les Annales De La Recherche Urbaine 57–58:212–217
Jounin N (2009) Chantier interdit au public. Enquête parmi les travailleurs du bâtiment. La Découverte
Latour B (2015) Face à Gaïa: Huit conférences sur le nouveau régime climatique. La Découverte
Lefrançois D (2014) Le parking dans le grand ensemble. Editions de la Villette
Lepoutre D (1997) Cœur de banlieue, Codes, rites et langages. Editions Odile Jacob
Mauger G (2019) Les jeunes des classes populaires sont des jeunes comme les autres. In: Masclet O, Misset S, Poullaouec T (eds) La France d'en bas? Idées reçues sur les classes populaires. Le Cavalier Bleu, pp 163–167
Ortar N, Subremon H (2018) L'énergie et ses usages domestiques. Anthropologie d'une transition en cours, Editions Petra
Ost F (2012) La nature hors la loi. L'écologie à l'épreuve du droit, La Découverte
Remy J (2015) L'espace, un objet central de la sociologie. Eres
Rosenvallon P (2020) Le parlement des invisibles. Le Seuil
Sennett R (1992) La ville a vue d'œil, Plon
Tarrius A (2002) La mondialisation par le bas. Les nouveaux nomades de l'économie souterraine, Balland
Vayssière B (1988) Reconstruction/déconstruction. Le hard french ou l'architecture française des trente glorieuses. Editions Picard
Zetlaoui J (2013)L'implication des habitants dans les projets d'écoquartiers en France. Vers des démarches intégrées.Cahiers Ramau n°6, pp 239–254

Chapter 13
Confusing the Spatial with the Social: Can Ethnography Offer a Way Out?

Jeanne Mosseray and Nele Aernouts

Abstract Urban renewal policies applied over Europe since the 1990s have been characterized by an integrated approach towards neighbourhood regeneration, combining an interplay of social and spatial intentions and strategies. In this contribution, we develop an ethnographic account of the occupation of the ground floors in the social high-rise estate of Peterbos, Brussels, in order to show the necessity for studying and translating such interplay locally. While over time, urban designers and spatial planners developed several proposals to include facilities in the plinths of the buildings in order to 'activate' the neighbourhood and make it livelier and more vibrant, we highlight such 'activation' by shopkeepers, social and community workers is less straightforward. The current occupancy of these spaces is characterized by the embodiment of the spaces by facility managers; their strategies, and those of their clients within and outside the spaces; and different forms of in- and exclusion. As such, the contribution highlights how an ethnographic approach could contribute to making more informed decisions on the design of such spaces.

Keywords Urban regeneration · Modernist estates · Ground floors · Ethnography

J. Mosseray (✉) · N. Aernouts
Department of Geography, Cosmopolis Centre for Urban Research, Vrije Universiteit Brussel, Brussel, Belgium
e-mail: jeanne.mosseray@vub.be

N. Aernouts
e-mail: nele.aernouts@vub.be

N. Aernouts et al. (eds.), *Urban Living Lab for Local Regeneration*,
The Urban Book Series, https://doi.org/10.1007/978-3-031-19748-2_13

13.1 Introduction: The Social Effects of Urban Regeneration

In response to the rather insensitive and undemocratic town planning approaches of post-war modernism,[1] urban regeneration policies of the last decades in Europe have been explicitly geared to achieving both social and spatial aims. A social effect is expected from physical measures (Van Kempen and Bolt 2009) through projects that cover issues such as unemployment, social cohesion, and education.

The Neighbourhood Contract programme has played a pioneering role in this (see, for instance, Berger 2019; Lenel 2014). It is applied in the most vulnerable areas of Europe, which combine high unemployment rates, low average incomes, and high density. In these territories, urban design and architecture projects are developed in partnership with local associations. Projects address public space, housing and public facilities, and socio-economic issues. Combining these interventions, the programme aims to socially revitalize neighbourhoods mainly through the transformation of the built environment. Such social revitalization is further supported by the development of participatory trajectories. Citizen participation in this programme is seen as a way to increase urban planners' understanding of an area and directly engage residents in the transformation of their surroundings. As such, citizen participation is a central element of the aim of social transformation.

While integrated and socially sensitive regeneration programmes are generally appreciated, the aims underlying them have not been unanimously welcomed. Many scholars have criticized urban transformation interventions aimed at, for instance, increasing social cohesion and social mix (Atkinson 2008; Blanc 2010; Blokland and Eijk 2010; Lenel 2013; Levin et al. 2014; Loopmans 2010; Van Kempen and Bolt 2009). Authors like Fijalkow (2017) have highlighted that pointing at the need for social cohesion overshadows local experiences and promotes the idea of a lack of social cohesion. It stigmatizes the current population by describing it in terms of problems. Expressing the need for a more social mix is even more stigmatizing, as it starts from the idea that a wealthier population would offer a solution to the 'lack' of social cohesion. While a considerable body of literature in the field of urban studies has been dedicated to the subject, to our knowledge less is known about the way this plays out in the context of the architecture and urban design practice. In this contribution, therefore, we show how regeneration intentions, combining social and spatial aims, are applied in the high-rise estate Peterbos in Brussels, Belgium. We do this by providing an ethnographic account of the occupation of the ground floor of the estate, which is currently undergoing regeneration. In Brussels, high-rise social estates are generally not subject to integrated and comprehensive planning policies that approach the neighbourhood as a whole. Their regeneration is tackled

[1] In the Brussels Capital Region, town planning activists coined the term 'Bruxellization' to refer to the laissez-faire approach of city authorities regarding urban planning, leaving the matter up to real estate developers. Spatially, this was reflected by drastic slum clearance programmes, with areas being replaced by high-rise office towers and housing.

case by case and only partially, according to the strategic plan[2] of the social housing company (in charge of the renovation of housing) and/or efforts[3] of the municipality (often in charge of the regeneration of public space). Neighbourhood Contracts were developed in these areas only recently, and for the first time in Peterbos.

Large-scale social estates are especially interesting for studying the dual aims of regeneration, as they are subject not only to interventions targeting social cohesion and social mix but also to projects aimed at crime reduction, improved neighbourhood reputation, and residential stability (see, for example, Arthurson 2012; Kleinhans 2004; Tunstall and Fenton 2006; Van Kempen and Bolt 2009).[4] Although theories on the connection between design and social effects have been widely criticized by scholars in the field, specific ideas continue to be part of regeneration intentions (Bolt 2018). This contribution will show how these rather speculative intentions influence architecture and urban design practice and hamper the development of a careful understanding of the actual use, occupation, and appropriation of spaces. By doing so, it hopes to highlight the potential merits of ethnography for developing a more subtle social reading of space.

13.2 An Ethnography of the Ground Floors in Peterbos

The social housing estate of Peterbos is located at the western fringe of the Brussels Capital Region and is composed of eighteen high- and mid-rise blocks. Eleven of the eighteen buildings host public facilities on their ground floors, while the others accommodate dwellings and technical or storage spaces. Since 2017, the neighbourhood has been subject to various renovation programmes. Some of these renovations are executed by social housing companies, which benefit from regional subsidies, while others are organized by the municipality, which was granted a Sustainable Neighbourhood Contract programme. This programme includes a master plan for the public space and the construction of public facilities. The ethnography of the occupation of the ground floors in Peterbos was developed in parallel with and independently from the renovation projects. However, being present in the neighbourhood from the start of the programme gave us a certain degree of influence on

[2] At the regional level, the *quadrennial investment plan* defines a framework for each company according to a strategic plan that they have to present to the regional agency.

[3] As the owner of the public space, the municipality can invest using the municipal budget or apply for different forms of funding at the European, national, and regional levels (Mosseray, Aernouts, & Ryckewaert, 2020).

[4] The assumption of a causal relationship between the built environment and human behaviour is not innocent. The architectural determinism of authors like Oscar Newman (1972)—who promoted concepts such as defensible space, situational prevention, and security design—inspired Alice Coleman (1985) to entirely denounce the heritage of large-scale social estates. It also supported politicians like Thatcher in their intention to privatize or demolish a substantial amount of social dwellings. Also, in France, the residentialization (the splitting up of large-scale social estates in autonomous residential units) of the social stock is seen as a cure for the social problems associated with it (Cupers, 2016).

them. This position permitted us to participate in various committees to discuss the project and, on some occasions, present our findings or reflections on the site. As the regeneration process is ongoing at the time of writing, it is difficult to measure whether or how our contribution will influence the final outcome. Our presentations and discussions did lead to some changes in the urban development process. They impacted the way the region conceives the renovation processes. Until now the renovation of the buildings was left to the housing companies, renovating one building (part) after the other. Throughout our discussions and due to our collaborations with the planning stakeholders, we were able to put the need for a broader study on the agenda, in order to develop a vision of the area as a whole.

Our ethnography in the neighbourhood of Peterbos was part of the SoHoLab action research project in Brussels and was mainly based on participant observation, for which one of the researchers lived in an apartment in the estate for twelve months. During these months, she also worked in a social services centre on the ground floor of one block. The ethnography was mostly conducted by the first author (further below named 'I'). Both authors ('we') engaged with the wider research framework (see also Aernouts et al., this book) and participated in discussions with governing bodies and urban designers.

Moving in provided me with the opportunity to participate in the daily life of the neighbourhood. As a resident, I was able to use the services on the ground floor as any other resident could. Some of these services were specifically dedicated to certain uses and people. Therefore, for each place, I adapted the method of observant participation in order to be able to regularly visit it as a 'legitimated' participant: becoming a visitor of the restaurant for the elderly, working one day a week at the social services desk as a volunteer, enrolling in a sewing course during the summer holidays, doing grocery shopping in the neighbourhood, using the local laundromat, and visiting the community centre. I regularly visited these different facilities and got to know all the social assistants, concierges, community workers, and technical staff working on the site. I accepted any invitations for a longer discussion. The service providers of the grocery shop, the municipal social desk, the community centre, the social restaurant, and the youth club became privileged witnesses in the neighbourhood. Apart from visiting the facilities, I experienced the neighbourhood spatially, taking walks in the area, meeting and visiting neighbours, observing the way people use public space, listening to conversations, and speaking with passers-by.

In participant observation, the main instrument of the ethnographer is their personality. It is, thanks to my personal relationships with service providers or users of these services, that I could observe, participate in, spatially experience, and discuss the ground floor services. I collected this data in my notebook. Furthermore, I recorded interviews with professionals and residents in order to trace their discourses and personal residential or working trajectories in the neighbourhood. This contribution is specifically based on four in-depth recorded interviews, organized after two years of exchange.

I developed tools to engage with the built form during the ethnographic fieldwork, including diverse exercises in order to obtain a detailed understanding of the built environment as an independent component. For example, the subject of ground floor occupation arrived very quickly in the fieldwork, as it is a delicate subject for the associations present in the neighbourhood. They want to be present and visible while they need enough space to support their collective activities. Therefore, spaces on the ground floor were continuously under pressure, being considered a potential place for facilities. In order to help the associations with their aims, I mapped all ground floor occupations. This enabled them to envisage the renovation of existing spaces or argue for the need for additional space in another block.

As attention to the ground floors of the blocks became more present in urban regeneration discussions, I began to focus on their spatial arrangement. I combined methods of participant observation and 3D sketching. Building on observations in the grocery shop of Emine, a local shopkeeper, I then systematically started to map all ground floor facilities. I conducted interviews with each service provider and drew the ground floor spaces and interior arrangements in 3D. These drawings formed the basis for analysing and projecting the way users and residents appropriated and adapted places throughout time. I also used photography to support my analysis and worked with a visual artist living in Peterbos to personify each individual story from the interview in a drawing.

13.3 The Ground floor's Place in the Regeneration Discourse

The design of the ground floors in Peterbos contains remnants of the CIAM modernist architecture movement. Some blocks are constructed on piles, which reinforce the idea of an infinite green carpet. In several buildings, modernist elements such as covered promenades, a colonnade, transparent spaces, and a rhythm of horizontal slabs can be found. In contrast to the initial design, the ground floors became rather functional spaces and less valued as patrimonial and architectural elements. They were either filled in to accommodate specific functions or closed down as a security solution. Facing a lack of places for facilities, local associations transformed technical spaces into workspaces. The ad hoc implementation of facilities and private dwellings over time resulted in a makeshift assembly of low ceilings, little daylight, bad acoustics, and dark spaces, blurring the architectural quality of the few CIAM principles that were applied in the neighbourhood (Fig. 13.1).

Rather than focusing on this loss of architectural quality, urban designers and architects turned their gaze to the occupation of the ground floor in the different renovation plans that were developed during our research in Peterbos. This interest resulted in a common agreement that the 'activation' of the ground floor was a crucial element of a successful renovation operation. It would provide a necessary transition between the public and the private sphere, deemed necessary in order to increase

Fig. 13.1 The architecture of the ground floors in Peterbos

social surveillance.[5] It also presented the possibility to expand new activities and facilities in order to support a functional mix. Further, it would, create landmarks and support a 'positive' appropriation.

The architects in charge of the design of the final master plan refined the Sustainable Neighbourhood Contract programme's proposal, suggesting a more careful reading of places yet still speaking in terms of 'activation' and 'new functions'. They envisaged commercial and economic activities to offer more facilities and services to local residents and potentially create job opportunities. Nevertheless, they faced several limitations that narrowed their ambitions and reflections. Social housing companies are not allowed to support non-residential uses in their buildings above a certain percentage. Next, the architects were asked to design a master plan for the public space that did not officially include the ground floors. The ethnography will illustrate how activating the ground floors faced not only legal and architectural challenges but should also overcome social obstacles.

13.4 The Neighbourhood Facilities: At the Nexus Between a Collective Mental Geography and Individualities

13.4.1 The Grocery Shop of Emine

I visited the grocery shop of Emine for the first time with my colleague. Emine was excited to have two 'international' visitors (being French and Dutch speaking, we communicate in English) in her shop and welcomed us with her best English. We were looking for some food in order to be able to continue our field visit, and she offered us a sandwich of her specialty. We naturally visited this shop instead of the two others, quite similar, present on the site, maybe due to its proximity to the playground. For us, it represented the most lively place in the afternoon.

---

[5] This idea was also coined in Newman's (1972) concept of defensible space: Architecture should be surrounded by a defensive area, which is provided by a gradient between public and private space. This would allow for social control from the dwellings and as a consequence prevent crime.

Some weeks later, I visited the shop as a 'resident' in order to buy bread. I began to recognize the faces of some people sitting in the shop, who were talking with the owner and other customers. During my first visits, I didn't dare ask for a coffee and sit there. Once she finally questioned me about my presence, Emine remembered my visit from a few weeks ago with a friend. From that moment onwards, we developed a close relationship: I passed by at least once a week and, depending on the crowd, chatted with her and other customers. She then referred to me as someone special whom she could confide.

Her shop is where Emine spends most of her time. Since 2013, she and her husband work there 7/7 from 9 to 9. Their life trajectories resulted in them investing in the shop as a common project. Previously, her sister had owned the shop, and before that, her cousin. This has contributed, since the 1990s, to the continued use of the name 'chez la Turcque' in the neighbourhood. Emine left her much-loved job (working as a lacemaker at the Grand Place in Brussels for about 15 years) after she got married in order to work with her husband, a newcomer from Turkey. As her husband did not master any of Belgium's national languages, she saw the shop as an opportunity to work together, applying the same strategy her sister and cousin applied previously with their Turkish husbands.

Contrary to her sister, Emine does not rent the apartment above the shop, which is attached to it. With her husband, she decided to stay in their newly renovated home 10 km from the area. She expresses the difficulty of commuting, leaving her with very little free time at home.

She took over the shop and was confident she would be able to manage the enterprise differently and better than her sister. The previous bankruptcy and bad experiences with burglaries did not scare Emine. However, she points to various difficulties involved in maintaining a shop like hers: being accepted by the neighbourhood, the threat of theft, coexisting with youth hanging around the place, the high expectations of her clients... As she invested her savings in this project, she does not have any alternative than persevere in the running of the shop, despite the difficulties of the neighbourhood. Recently, however, after seven years, Emine and her husband have decided to hand over the shop, as the difficulties became too significant.

In terms of space, Emine struggles to find the right layout: to be visible enough, respect the rules of the landlord, manage her stock, and organize her space as protectively as possible and yet be welcoming. She describes her shop in terms of protection first: she created a shield with shutters, window protection, and cameras. Her counter fences in half the merchandise. From there, she is able to see outside without being seen too much. Her position next to the playground brings in many children buying sweets, an aspect of her job she likes (Fig. 13.2). However, her central position in the neighbourhood, together with the 'protective' adaptations, sometimes turns out to be at her disadvantage. Her shop is often appropriated and invaded by a group of youngsters, who use it to shield themselves from the police. She is therefore in constant negotiation with them.

Emine also mentions the necessity for her to adapt her offer to specific needs, for example by acting as a nearby grocery shop for physically disabled people who represent an important part of her clients or by developing a partnership with residents

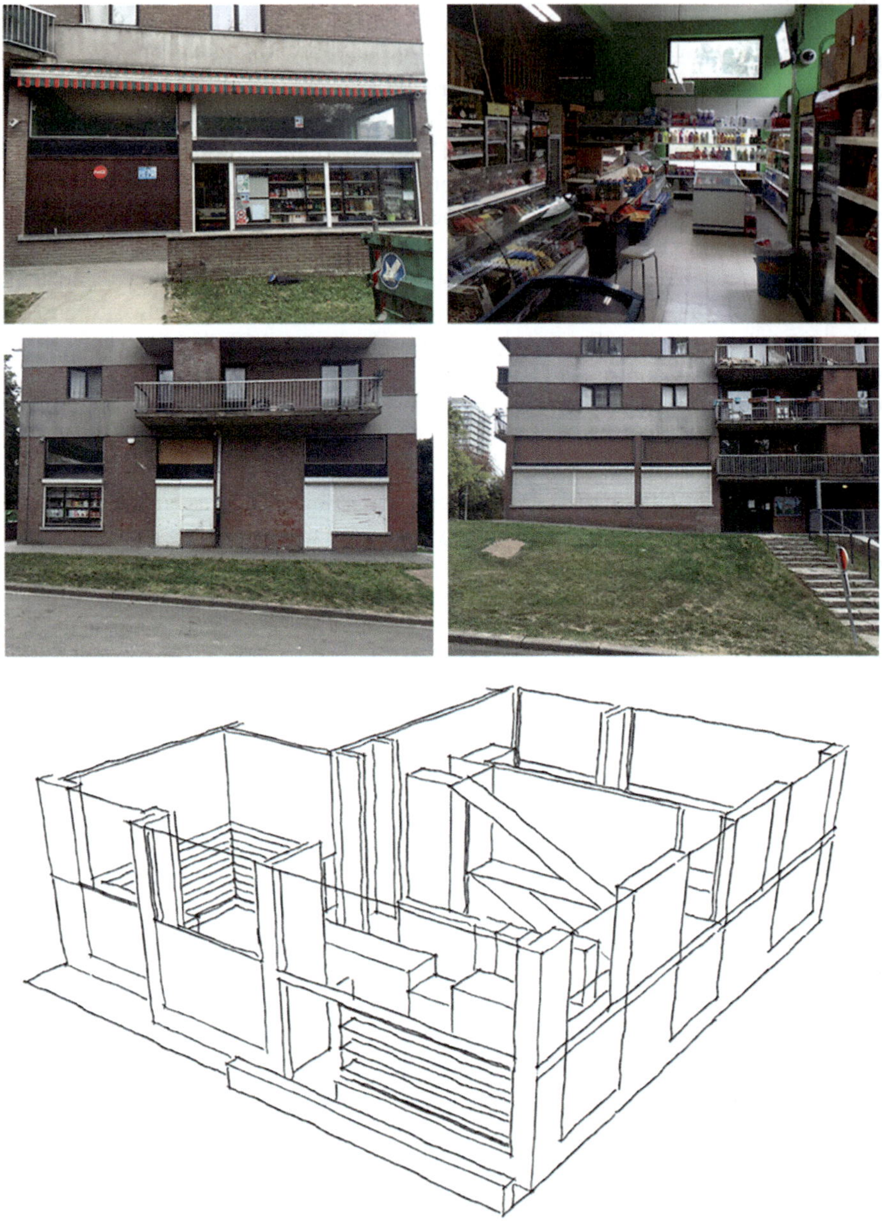

Fig. 13.2 The shop of Emine

who sell their homemade dishes. These adaptations not only result from the socio-economic situation of residents but also from the particular needs they face. Emine therefore sells specific products (bottles of water that are too heavy for pedestrians to carry a long way) and offers specific services (hot coffee, sandwiches, home delivery, and the fulfilment of individual requests).

13.4.2 Facilities as Points of Reference

The deep personal involvement highlighted in the story of Emine reflects a common trend in Peterbos. In my interviews with the managers of the different facilities in the neighbourhood—the shopkeepers, community workers, municipal employees, and volunteers (largely women)—it became clear that their individual investment often results in embodying their service (Fig. 13.3). Residents refer to '*chez Françoise*', '*chez Khadija*', '*le groupe de Youssef*', '*le local de Pierre*', '*chez la Turque/ chez Emine*', and '*avec Julie*' to describe the municipal social services desk, the social restaurant, the youth club, etc. These people's individual life stories, together with their long-term presence in the neighbourhood, contribute to making them common figures and references. Youssef grew up in Peterbos. As a young man, he was looking for a space to do activities with his friends. Françoise, the initiator of the social desk, explains her long attachment to the municipality and the neighbourhood. She encountered many difficulties in finding colleagues and worries about her succession in the future. Newer users of the ground floors, such as Emine, express the necessity to be accepted by the neighbourhood, given the long engagement of others. Most managers entered the estate due to their life trajectories: being a former or current resident, having an immigrant background, being attached to their municipality… Being 'on the first line', they are often confronted with difficult life circumstances, such as loneliness, poverty, and disease. The trajectories and their understanding of these conditions contribute to their deep involvement and enable the preservation of facilities in the neighbourhood.

Fig. 13.3 Artistic interpretation of four facility managers (©Frederic Chapelle)

13.4.3 Facilities as Chameleons

The 'in-between' position of facility managers, who are physically present but do not experience residents' difficulties, makes them engage with the spatial layout of the neighbourhood in a specific way. Some architectural elements support their occupation, while other elements invoke uses that they do not desire. A colonnade allows the pharmacy to display lettering and the Pakistani manager of the grocery shop to cover his fruit and vegetables outside. The location of Emine's shop enables her to attract some clients while it discourages others, who are afraid of experiencing undesirable situations. The low wall in front of her shop invites youth to hang around, which makes some people feel uncomfortable. Also, Françoise, the social assistant that manages the social desk near Emine's shop, appreciates her central position in the neighbourhood as a way to welcome as many residents as possible. For the same reason, the community workers located in the extreme north of the neighbourhood, which is less frequented by passers-by, moved their community space in order to attract more residents to their activities. Yet, the central position of Françoise sometimes makes her feel trapped, for instance when she is exposed to aggression in front of her window. She feels that the transparency of her desk is detrimental to the intimacy she would like to offer to her visitors. Likewise, once they moved, the community workers decided to relocate their offices to a more remote location in a former dwelling. The location is less transparent, allowing them to be less visible and keep their distance from regular clients with psychological problems.

The location and spatial layout of the facilities have obliged facility managers to develop adjustment strategies. The story of Emine shows how the geographic location and architecture of the shop lead her to engage with its spatial layout in a certain manner, resulting in a specific architecture and aesthetic. Her adaptations are in line with the opportunities offered by high-rise buildings and with the legal status of renting, as they depend on the rules and expectations of the landlord. In the case of the shop, many windows have been sealed by shutters in order to prevent vandalism, creating an image of an unused or closed space. Initially, Emine wanted to put advertising on the closed shutters, but this was not allowed by the landlord. She also decided to diminish the surface of the shop to reduce the cost of rent and the risk of theft, while improving the view of her merchandise. These practices show how activating and giving life to the surroundings cannot be reduced to the presence of facilities.

More generally, each of the facilities present in the eleven buildings today reveals different kinds of adaptations (Fig. 13.4). These adaptations—the shutters of Emine, curtains of the sewing course, drawings on the window of the computer class, a reorganized public terrace of the community centre—make the facilities less identifiable to people that do not live in the neighbourhood. Residents develop their mental maps differently. Over time and depending on their needs, they have learnt what the neighbourhood has to offer. Within the neighbourhood, residents use spatiotemporal references—through shared terms such as '*chez la Turc*', '*à la plaine*', '*au riso*', and '*le parc du pont*'—to refer to the facilities offered there. Behind these names, they see the services the area offers, and they orient themselves to these accordingly: where to find administrative help and at what time, where to buy a sandwich without any apparent sandwich bar being present in the neighbourhood, or where to get a free coffee.

Consequently, facilities and their services are often hidden from the general public. Recognizing the activities taking place behind these shutters or small doors requires a knowledge of the shop's interior and of implicit rules. Emine, for instance, hides certain activities because she might be at the limit of legality by offering these services, but also because they are not offered to everyone. In her case, only women are invited to sit to have their coffee, while men will be offered takeaway coffee only.

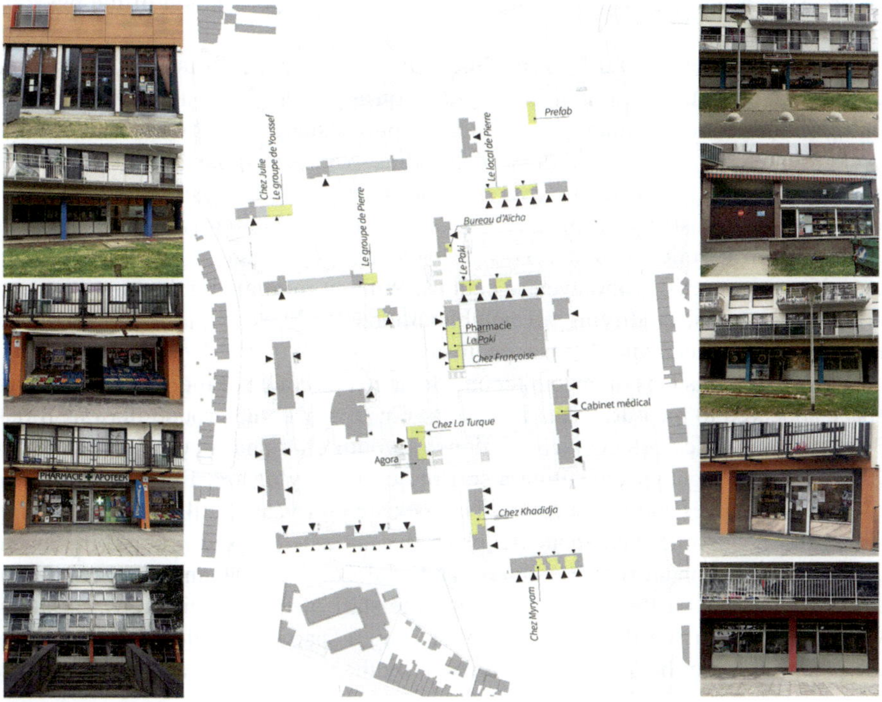

Fig. 13.4 Peterbos ground and public facilities—Facades of ground floor facilities (community centre—computer classroom—grocery shop—pharmacy—social restaurant—grocery shop—grocery shop—social housing company desk—social desk—sewing classroom)

13.4.4 Facilities as Spaces of Encounter and Coexistence

Emine's story illustrates how ground floor facilities can become places of encounters, mixing generations, origins, or genders while also welcoming visitors to the neighbourhood, from social workers to clients of drug dealers to children. The location, interior of the shop, and Emine's personality add to this. However, these encounters are experienced in manifold senses, as places of not only exchange and meeting, but also conflict and confrontation. Broadly considered positive, supporting sparks of social cohesion, the facility also generates uncomfortable feelings. As the shop front is from time to time the scene of violent exchanges between municipal workers and youngsters hanging around there, some residents avoid it as much as possible. These ad hoc scenes and consequences influence the overall appreciation of the place by residents.

Moreover, other ground floor facilities, such as the social restaurant, the sewing course, and the youth club, are meant to function as places of encounter, welcoming a mixed public. In reality, they accommodate a public sharing the same ethno-cultural background (Fig. 13.5). The story of Khadija, who is responsible for the senior restaurant, highlights this. She expresses her pain when discussing the very racist welcome the exclusively European (Belgian, Spanish, and Italian) restaurant regulars once gave to a veiled woman. This de facto categorization of facilities leads to unwelcoming reactions and even rejections. Without minimizing the importance of non-mixed places, qualifying all neighbourhood facilities as places of encounter needs to be nuanced from this perspective.

Building on this lens of encounter and diversity, the value of more flexible uses of space appeared to me during the fieldwork. For instance, the ground floor of Block 3 historically welcomes two associations (a youth club and a community organization), which together constitute a sort of community centre. Both projects have different objectives and target groups but coexisted for years until 2020, embodying a successful, flexible, and inclusive place that attracted mixed groups (mainly in terms of generational diversity: elderly and adults on one side and youth, children, and their parents on the other). Although the staff of the organizations got along well, they did not see the advantage of sharing a space. They decided to split, as the place could not be appropriated enough by the young public and had too many

Fig. 13.5 The sewing course and the social restaurant

logistic challenges. Although this and other places, such as the social restaurant, were implemented with the goal of offering an open and flexible space, a series of decisions led to the allocation of spaces to a specific organization catering to specific groups, excluding others.

13.5 A Potential Role for Ethnography in Offering a More Subtle Social-Spatial Reading?

What can we learn from an ethnography of space? Could it help us develop more socially sensitive accounts of spatial interventions, instead of giving social justifications for spatial design? In this contribution, we discussed three dimensions that characterize facilities and that, according to us, need careful consideration when rethinking the ground floor spaces: the embodiment of spaces by facility managers; the way these facility managers and their clients engage with space; and the forms of in- and exclusion that accompany this.

First, 'activating' ground floors preferably builds on a reflection on the kinds of activities that may take place. In a modernist typology, historical building blocks that can be used to read and orient oneself in the city do not apply. In Peterbos, ground floor activities became important landmarks. Not necessarily the spaces in and of themselves, but rather the offer in and embodiment of the spaces by facility managers create spatiotemporal references. We see that this role of facility managers is strongly intertwined with their personal history and attachment to the neighbourhood. Still, they continue to encounter several challenges when becoming (and remaining) facility managers, which makes such activation of spaces not obvious.

Second, a call for activation requests an understanding of how facility managers and their clients engage with space. In urban design, the activation and appropriation of places are often associated with a spatial interface between the private and the public, between the street and the building. As such a link is not obvious in high-rise estates, spatial professionals may want to generate activity on the ground floor in order to support this kind of appropriation. This is in line with the idea that a mixed-use neighbourhood is the desired pattern of physical development that can achieve a more vital, vibrant, attractive, safe, viable, and sustainable urban lifestyle (Mehta 2009). The layout of the shops in Peterbos offers some interesting insights into the nature of the transition. While architectural components such as a colonnade support the display of goods and personalizing of shop fronts, other elements, such as a low wall, are not necessarily seen as desirable, as they attract loitering teens. Many glass facades have been partially covered (e.g. from totally open to totally closed) in reaction to their exposure to large public spaces. Inside, the decoration of interiors by facility managers builds on a transition within the space, with some areas of the facility being more open and public and others more closed, while still keeping an eye on the more public spaces. Next to these material adaptations by facility managers meant to discourage or attract clients, we also identified immaterial strategies, such as offering services to people whom facility managers could identify. These material and immaterial components and strategies create different gradients of intensity and attractiveness and punctuate life in the neighbourhood.

Third, in the design process in Peterbos, ground floors were defined as places to be equipped in order to support appropriation, encounters, and potential social cohesion. However, residents have an ambivalent attachment to these places. Some facilities generate activity, while others are avoided by residents. Depending on the service provider, the offer, and the location, we can distinguish three gradients of encounter: between diverse groups, such as in shops, where loose communities are constructed; within certain groups, mostly through social or community organizations; and between and within groups at the same time, in multifunctional spaces that invite diverse groups in one location.

Ultimately, spatial design remains the material outcome and result of construction and renovation processes. The social realities and social constructions already at stake can support the reimagination of this design, but the interpretation and spatial translation of this reading remain in the hands of the designer. Succeeding in providing for the diverse spatial concerns of the groups exploiting, using, and occupying it is the first step towards a more socially sensitive urban design. The images supporting this contribution show that this might start not only with a careful reading of a space but also with experimenting with new forms of expression and representation.

References

Arthurson K (2012) Social mix and the city: challenging the mixed communities consensus in housing and urban planning policies. CSIRO Publishing

Atkinson R (2008) European urban policies and the neighbourhood: an overview. Proceedings of the Institution of Civil engineers—urban design and planning, 161(3), 115–122. https://doi.org/10.1680/udap.2008.161.3.115

Coleman A (1985) Utopia on trial. Vision and reality in planned housing. Hilary Shipman, London

Berger M (2019) L'invention du contrat de quartier. In Le temps d'une politique: Chronique des Contrats de quartier bruxellois. CIVA, pp 29–44

Blanc M (2010) Métiers et professions de l'urbanisme : l'ingénieur, l'architecte et les autres. Espaces Et Societes 142(2):131–150

Blokland T, Eijk G (2010) Do people who like diversity practice diversity in neighbourhood life? Neighbourhood use and the social networks of 'diversity-seekers' in a mixed neighbourhood in the Netherlands. J Ethn Migr Stud 36(2):313–332. https://doi.org/10.1080/13691830903387436

Bolt G (2018) Who is to blame for the decline of large housing estates? An exploration of socio-demographic and ethnic change. In Hess DB, Tammaru TT, van Ham M (Eds), Housing estates in Europe: Poverty, ethnic segregation and policy challenges. Springer International Publishing, pp 57–74. https://doi.org/10.1007/978-3-319-92813-5

Cupers K (2016) Mapping and making community in the postwar European city. J Urban Hist 42(6):1009–1028. https://doi.org/10.1177/0096144216675044

Fijalkow Y (Ed). (2017) Dire la ville c'est faire la ville. La performativité des discours sur l'espace urbain. Septentrion

Kleinhans R (2004) Social implications of housing diversification in urban renewal: a review of recent literature. J Housing Built Environ 19(4):367–390. https://doi.org/10.1007/s10901-004-3041-5

Lenel E (2013) Social mix in public urban action in Brussels: Project or political language? Brussels Studies, 65 (online). https://doi.org/10.4000/brussels.1134
Lenel E (2014) L'espace public 'revitalisé' comme gouvernance de l'intime: Le cas d'une salle de sport aménagée dans le cadre d'un dispositif public bruxellois. Ambiances, July (online). https://doi.org/10.4000/ambiances.441
Levin I, Arthurson K, Ziersch A (2014) Social mix and the role of design: competing interests in the Carlton public housing estate redevelopment, Melbourne. Cities 40:23–31. https://doi.org/10.1016/j.cities.2014.04.002
Loopmans M (2010) Threatened or empowered? The impact of neighborhood context on community involvement in Antwerp. Belgium. Urban Affairs Review 45(6):797–820. https://doi.org/10.1177/1078087410364604
Mehta V (2009) Look closely and you will see, listen carefully and you will hear: Urban design and social interaction on streets. J Urban Des 14(1):29–64. https://doi.org/10.1080/13574800802452658
Mosseray J, Aernouts N, Ryckewaert M (2020) Brussels. In: Aernouts N, Maranghi E, Ryckewaert M (Eds), The regeneration of large-scale social housing estates. Spatial, territorial, institutional and planning dimensions (pp 133–185). SoHoLab.
Newman O (1972) Defensible space: crime prevention through urban design. Macmillan, New York
Tunstall R, Fenton A (2006) In the mix, a review of mixed income, mixed tenure and mixed communities. Housing Corporation, Joseph Rowntree Foundation & English Partnerships
Van Kempen R, Bolt G (2009) Social cohesion, social mix, and urban policies in the Netherlands. J Housing Built Environ 24(4):457–475. https://doi.org/10.1007/s10901-009-9161-1

GPSR Compliance

The European Union's (EU) General Product Safety Regulation (GPSR) is a set of rules that requires consumer products to be safe and our obligations to ensure this.

If you have any concerns about our products, you can contact us on ProductSafety@springernature.com

In case Publisher is established outside the EU, the EU authorized representative is:

Springer Nature Customer Service Center GmbH
Europaplatz 3
69115 Heidelberg, Germany

Batch number: 10844706

Printed by Printforce, the Netherlands